APPLICATIONS OF POLYMERS AND PLASTICS IN MEDICAL DEVICES

PLASTICS DESIGN LIBRARY (PDL) PDL HANDBOOK SERIES
Series Editor: Sina Ebnesajjad, PhD (sina@FluoroConsultants.com) President, FluoroConsultants Group, LLC Chadds Ford, PA, USA http://www.FluoroConsultants.com

The **PDL Handbook Series** is aimed at a wide range of engineers and other professionals working in the plastics industry, and related sectors using plastics and adhesives.

PDL is a series of data books, reference works and practical guides covering plastics engineering, applications, processing, and manufacturing, and applied aspects of polymer science, elastomers and adhesives.

Recent titles in the series
Polymer Hybrid Materials and Nanocomposites, Saleh (ISBN: 9780128132944)
Design and Manufacturing of Plastics Products, Pouzada (ISBN: 9780128197752)
Automotive Plastics and Composites, Greene (ISBN: 9780128180082)
The Effect of Long Term Thermal Exposure on Plastics and Elastomers, McKeen (ISBN: 9780323854368)
Introduction to Fluoropolymers, Ebnesajjad (ISBN: 9780128191231)
The Effect of Radiation on Properties of Polymers, McKeen (ISBN: 9780128197295)
Service Life Prediction of Polymers and Coatings, White (ISBN: 9780128183670)
A Practical Guide to Plastics Sustainability, Biron (ISBN: 9780128215395)
Applications of Fluoropolymer Films, Drobny (ISBN: 9780128161289)
Recycling of Flexible Plastic Packaging 1, Niaounakis (ISBN: 9780128163351)
Plasticizers Derived from Post-Consumer PET 1, Langer (ISBN: 9780323462006)
Polylactic Acid 2, Sin (ISBN: 9780128144725)
Durability and Reliability of Polymers and Other Materials in Photovoltaic Modules, Yang, French & Bruckman
    (ISBN: 9780128115459)
Fluoropolymer Additives 2, Ebnesajjad & Morgan (ISBN: 9780128137840)
The Effect of UV Light and Weather on Plastics and Elastomers 4, McKeen (ISBN: 9780128164570)
PEEK Biomaterials Handbook 2, Kurtz (ISBN: 9780128125243)
Hydraulic Rubber Dam, Thomas et al. (ISBN: 9780128122105)
Electrical Conductivity in Polymer-based Composites, Taherian & Kausar (ISBN: 9780128125410)
Plastics to Energy, Al-Salem (ISBN: 9780128131404)
Recycling of Polyethylene Terephthalate Bottles, Thomas et al. (ISBN: 9780128113615)
Dielectric Polymer Materials for High-Density Energy Storage, Dang (ISBN: 9780128132159)
Thermoplastics and Thermoplastic Composites, Biron (ISBN: 9780081025017)
Recycling of Polyurethane Foams, Thomas et al. (ISBN: 9780323511339)
Introduction to Plastics Engineering, Shrivastava (ISBN: 9780323395007)
Chemical Resistance of Thermosets, Baur, Ruhrberg & Woishnis (ISBN: 9780128144800)
Phthalonitrile Resins and Composites, Derradji, Jun & Wenbin (ISBN: 9780128129661)
The Effect of Sterilization Methods on Plastics and Elastomers, 4e, McKeen (ISBN: 9780128145111)
Polymeric Foams Structure-Property-Performance, Obi (ISBN: 9781455777556)
Technology and Applications of Polymers Derived from Biomass, Ashter (ISBN: 9780323511155)
Fluoropolymer Applications in the Chemical Processing Industries, 2e, Ebnesajjad & Khaladkar
    (ISBN: 9780323447164)
Reactive Polymers, 3e, Fink (ISBN: 9780128145098)
Service Life Prediction of Polymers and Plastics Exposed to Outdoor Weathering, White, White & Pickett,
    (ISBN: 9780323497763)
Polylactide Foams, Nofar & Park (ISBN: 9780128139912)
Designing Successful Products with Plastics, Maclean-Blevins (ISBN: 9780323445016)
Waste Management of Marine Plastics Debris, Niaounakis, (ISBN: 9780323443548)
Film Properties of Plastics and Elastomers, 4e, McKeen, (ISBN: 9780128132920)
Anticorrosive Rubber Lining, Chandrasekaran (ISBN: 9780323443715)
Shape-Memory Polymer Device Design Safranski & Griffis, (ISBN: 9780323777973)
A Guide to the Manufacture, Performance, and Potential of Plastics in Agriculture, Orzolek, (ISBN: 9780081021705)
Plastics in Medical Devices for Cardiovascular Applications, Padsalgikar, (ISBN: 9780323358859)
Industrial Applications of Renewable Plastics, Biron (ISBN: 9780323480659)
Permeability Properties of Plastics and Elastomers, 4e, McKeen, (ISBN: 9780323508599)
Expanded PTFE Applications Handbook, Ebnesajjad (ISBN: 9781437778557)
Applied Plastics Engineering Handbook, 2e, Kutz (ISBN: 9780323390408)
Modification of Polymer Properties, Jasso-Gastinel & Kenny (ISBN: 9780323443531)
The Science and Technology of Flexible Packaging, Morris (ISBN: 9780323242738)
Stretch Blow Molding, 3e, Brandau (ISBN: 9780323461771)
Chemical Resistance of Engineering Thermoplastics, Baur, Ruhrberg & Woishnis (ISBN: 9780323473576)
Chemical Resistance of Commodity Thermoplastics, Baur, Ruhrberg & Woishnis (ISBN: 9780323473583)

To submit a new book proposal for the series, or place an order, please contact Brian Guerin, Acquisitions Editor at
    b.guerin@elsevier.com

APPLICATIONS OF POLYMERS AND PLASTICS IN MEDICAL DEVICES

Design, Manufacture, and Performance

SYED ALI ASHTER
Advanced Plastics Enterprise LLC., Corona, CA, United States

William Andrew is an imprint of Elsevier
The Boulevard, Langford Lane, Kidlington, Oxford, OX5 1GB, United Kingdom
50 Hampshire Street, 5th Floor, Cambridge, MA 02139, United States

Copyright © 2022 Elsevier Inc. All rights reserved.

No part of this publication may be reproduced or transmitted in any form or by any means, electronic or mechanical, including photocopying, recording, or any information storage and retrieval system, without permission in writing from the publisher. Details on how to seek permission, further information about the Publisher's permissions policies and our arrangements with organizations such as the Copyright Clearance Center and the Copyright Licensing Agency, can be found at our website: www.elsevier.com/permissions.

This book and the individual contributions contained in it are protected under copyright by the Publisher (other than as may be noted herein).

Notices
Knowledge and best practice in this field are constantly changing. As new research and experience broaden our understanding, changes in research methods, professional practices, or medical treatment may become necessary.

Practitioners and researchers must always rely on their own experience and knowledge in evaluating and using any information, methods, compounds, or experiments described herein. In using such information or methods they should be mindful of their own safety and the safety of others, including parties for whom they have a professional responsibility.

To the fullest extent of the law, neither the Publisher nor the authors, contributors, or editors, assume any liability for any injury and/or damage to persons or property as a matter of products liability, negligence or otherwise, or from any use or operation of any methods, products, instructions, or ideas contained in the material herein.

Library of Congress Cataloging-in-Publication Data
A catalog record for this book is available from the Library of Congress

British Library Cataloguing-in-Publication Data
A catalogue record for this book is available from the British Library

ISBN: 978-0-12-820980-6

For information on all William Andrew publications visit our website at https://www.elsevier.com/books-and-journals

Publisher: Matthew Deans
Acquisitions Editor: Brian Guerin
Editorial Project Manager: Fernanda A. Oliveira
Production Project Manager: Prem Kumar Kaliamoorthi
Cover Designer: Greg Harris

Typeset by TNQ Technologies

This book is dedicated to my alma mater, Aligarh Muslim University, and its Founder, Sir Syed Ahmad Khan.

"Jo Abr Yahan Se Utthega, Wo Sarey Jahan Par Barsega"

Contents

Biography   *xi*
Preface   *xiii*
Acknowledgments   *xv*

1. **Introduction to polymers and plastics for medical devices**   1
   1.1 History of medical devices   2
   1.2 What are medical devices   2
   1.3 Basic definitions   4
   1.4 Material used in medical devices   7
   1.5 Importance of plastics in medical devices   11
   1.6 General requirements of materials to qualify for medical devices   13
   1.7 Global nature of medical devices   15
   1.8 Socioeconomic factors   16
   References   22

2. **Classification of medical devices**   27
   2.1 Role of Food and Drug Administration   27
   2.2 Classification of medical devices   28
   2.3 Why are devices classified?   33
   2.4 Device classification panels   33
   2.5 Exemptions   33
   2.6 Emergency use authorization   35
   2.7 Premarket notification, 510(k)   36
   2.8 Premarket approval   36
   2.9 Postapproval market requirements   38
   2.10 FDA's Accelerated Approval Program   40
   2.11 FDA's Breakthrough Devices Program   41
   References   42

3. **Selection of materials for construction of medical devices**   45
   3.1 Overview of materials   45
   3.2 Material requirements for medical device qualification   49
   References   63

4. **Classification of plastics and elastomers used in medical devices**   65
   4.1 Performance-based selection of plastics and elastomers for medical devices   66

4.2	Plastics and elastomers used in FDA class I devices	70
References		77

5. Low performance demand plastics and elastomers for medical devices — 79

5.1	Introduction	79
5.2	Polypropylene	80
5.3	Polyethylene	82
5.4	Polyvinyl chloride (PVC)	85
5.5	Polystyrene	87
References		89

6. Medium-performance demand plastics and elastomers for medical devices — 91

6.1	Introduction	91
6.2	Polycarbonate	92
6.3	Polymethyl methacrylate	95
6.4	Polybutylene terephthalate	97
6.5	Polyphenylene oxide	98
6.6	Acrylonitrile butadiene styrene	100
References		102

7. High-performance demand plastics and elastomers for medical devices — 105

7.1	Introduction	106
7.2	Acetal copolymer	107
7.3	Polyetheretherketone	109
7.4	Polyphenyl Sulfone	112
7.5	Polysulfone	113
7.6	Polyphenylene sulfide	115
7.7	Polyvinylidene fluoride	117
7.8	Polyetherimide	119
7.9	Polydimethylsiloxane	121
7.10	Thermoplastic polyurethane	122
7.11	Thermoplastic elastomer	124
References		126

8. Plastics fabrication techniques — 129

8.1	Machining of plastics	129
8.2	Bonding	139
8.3	Staking	141
8.4	Two-part molding (silicone molding)	143
References		144

9. Plastic medical device manufacturing processes — 147
    9.1 Introduction — 147
    9.2 Extrusion — 148
    9.3 Injection molding process — 151
    9.4 Catheter manufacturing process — 153
    9.5 Other secondary operations — 158
    References — 162

10. Therapeutic applications of medical devices — 163
    10.1 Biosurgery — 164
    10.2 Cardiovascular heart rhythm — 169
    10.3 Vascular surgery — 178
    10.4 Interventional cardiology — 185
    10.5 Thoracic drainage systems — 191
    10.6 Prosthetic implants — 193
    10.7 In vitro diagnostics — 197
    References — 203

11. Applications of biobased polymers in medical devices — 209
    11.1 Introduction — 209
    11.2 Implantable medical devices — 210
    11.3 Diagnostics and labware devices — 218
    11.4 Healthcare electronic devices — 218
    11.5 Medical mask and medical tubing — 219
    References — 219

12. Regulatory requirements for medical devices — 223
    12.1 Overview — 223
    12.2 The Food and Drug Administration — 224
    12.3 European Union Commission — 233
    12.4 Other global agencies — 234
    12.5 International Medical Device Regulators Forum — 238
    References — 239

13. Market trends and global sourcing of medical devices — 241
    13.1 Market trends toward increased safety and quality of medical devices — 241
    13.2 Market trends in the medical device industry — 243
    13.3 Global sourcing of plastics — 245
    References — 248

14. Reprocessing of reusable medical devices — 251
- 14.1 Overview—reusable medical devices — 251
- 14.2 FDA guidance on reprocessing — 252
- 14.3 Factors affecting quality of reprocessing — 253
- 14.4 Cleaning validation—reusable medical devices — 255
- 14.5 Validation of the final microbicidal process — 256
- 14.6 Reusable medical device labeling — 256
- 14.7 Reprocessing of single-use medical devices — 258
- References — 259

15. Economics and future direction of medical devices — 261
- 15.1 Understanding the economics of medical device cost — 261
- 15.2 Migration of device manufacturing, validation and use — 270
- References — 274

Index — *277*

Biography

Dr. Syed Ali Ashter is the Founder and CEO at Advanced Plastics Enterprise LLC., a material evaluation, product development and quality compliance consulting company, which he founded in 2020. Dr. Ashter received his Bachelors (1998) from Aligarh Muslim University, India, in chemical engineering, M.S. (2002) and Ph.D. (2008) from University of Massachusetts Lowell in plastics engineering. Dr. Ashter worked as a postdoctoral research fellow at McMaster University, Hamilton, Canada, on next-generation formable films. Dr. Ashter has extensive experience working in medical device and biopharmaceutical industry developing products that saves patient lives.

Dr. Ashter is a member of Society of Plastics Engineers (SPE) and American Society for Testing Materials (ASTM). He has served on the Board of Directors for the SPE's Medical Plastics Division (MPD) since 2012. During his time on the Board, Dr. Ashter has held numerous roles within the Executive Committee. Currently, he is serving as the chair of the Medical Plastics Division since 2020.

Dr. Ashter has authored numerous conference papers and peer-reviewed journals. He has published three volumes in the Plastics Design Library series, Thermoforming of Single and Multilayer Laminates (2013), Introduction to Bioplastics Engineering (2016), and Technology and Applications of Polymers Derived from Biomass (2017).

Preface

Applications of Polymers and Plastics in Medical Devices: Design, Manufacture, and Performance is a comprehensive guide to plastic materials for medical devices, covering fundamentals, materials, applications, and regulatory requirements. This is an essential resource for engineers, R&D, and other professionals working on plastics for medical devices and those in the plastics industry, medical device manufacturing, pharmaceuticals, packaging, and biotechnology. In an academic setting, this book is of interest to researchers and advanced students in medical plastics, plastics engineering, polymer science, mechanical engineering, chemical engineering, biomedical engineering, and materials science.

The book is comprised of 15 chapters covering wide array of topics. Chapter 1 explores the history of medical devices, basic definitions, importance of plastics in medical devices, and impact of healthcare cost on patients. Classification of medical devices and the role that the Food and Drug Administration plays in regulating them is described in Chapter 2. Chapter 3 provides an overview on materials used in medical devices and their requirements for commercialization.

Materials such as plastics and elastomers play a significant role in revolutionizing healthcare. Selection of plastics and elastomers for medical device as well as identifying plastics and elastomers used in FDA class 1 devices is discussed in Chapter 4. Chapter 5, 6, and 7 explores properties and applications of low, medium, and high performance demand polymers. Information on commercially available medical grades and the supplier for each material is also discussed.

Chapter 8 discusses different plastics machining methods such as CNC machining, Plasma Cutters, Electric Discharge Machining, Turning, and Water Jet Cutting. Additionally, other operations such as bonding, stacking, and molding will also be highlighted. Medical device manufacturing includes all aspects of the fabrication of a medical device, from designing a manufacturing process to scale-up to ongoing process improvements. Chapter 9 discusses extrusion and injection molding processes with respect to medical device manufacturing. In addition, catheter manufacturing process will also be discussed.

Chapter 10 reviews some of the commercially available medical devices with applications in Biosurgery, Cardiovascular and Heart Rhythm,

Vascular Surgery, Interventional Cardiology, Thoracic Drainage Systems, Prosthetics Implants, and In Vitro Diagnostics. Chapter 11 discusses applications of bio-based polymers in medical devices focusing on four different areas: (a) implantable medical devices, (b) diagnostic and labware, (c) healthcare electronic devices, and (d) masks and medical tubing.

Every medical device manufacturer is required to comply with regulations of the country devices that are intended for use. Chapter 12 reviews the role of Food and Drug Administration (FDA) and other global agencies in regulating medical devices and ensuring medical devices are safe for patients. Chapter 13 discusses market trends toward safety and quality of medical devices and global sourcing of plastics. Chapter 14 provides FDA guidance on reprocessing, factor affecting quality of reprocessing, and cleaning validation of reusable medical devices. Chapter 15 reviews economics of medical device cost and migration of device manufacturing, validation, and use.

Acknowledgments

I would like to thank all corporate partners for being part of this book by providing copyright images of their product and equipment. I would also like to express my sincere gratitude to Sina Ebnesajjad, Editor for Plastics Design Library for his trust and continued support. Finally, I would like to thank my family, Tahira, Zayn, and Noor for their patience and unconditional support.

CHAPTER ONE

Introduction to polymers and plastics for medical devices

Contents

1.1 History of medical devices	2
1.2 What are medical devices	2
1.3 Basic definitions	4
1.4 Material used in medical devices	7
1.4.1 Metals	7
1.4.2 Polymers	9
1.5 Importance of plastics in medical devices	11
1.5.1 Sterilization	12
1.5.2 Safety	12
1.5.3 Comfort and ease-of-use	12
1.5.4 Design flexibility and applications	13
1.5.5 Cost	13
1.6 General requirements of materials to qualify for medical devices	13
1.6.1 Step 1—preliminary assessment	14
1.6.2 Step 2—screening	15
1.6.3 Step 3—rank	15
1.7 Global nature of medical devices	15
1.8 Socioeconomic factors	16
1.8.1 Graying populations require more medical care	16
1.8.2 Cost of care	17
1.8.3 Proliferation of procedures that require more medical care	18
1.8.4 Device design to reduce skill requirements for use	18
1.8.4.1 User system	*19*
1.8.4.2 Device system	*19*
1.8.5 Role of polymers in cost reduction and increasing the availability of devices	20
1.8.5.1 Material costs or processability	*20*
1.8.5.2 Product weight	*21*
1.8.5.3 Materials of construction	*21*
1.8.5.4 New technologies	*22*
References	22

Applications of Polymers and Plastics in Medical Devices
ISBN: 978-0-12-820980-6
https://doi.org/10.1016/B978-0-12-820980-6.00008-4

© 2022 Elsevier Inc.
All rights reserved.

1.1 History of medical devices

Medical devices have been around for centuries; however, significant growth has happened in the past 20 years. Some examples include development of heart and lung machines and identification of medical procedures for performing advanced brain surgery. Although the first application of medical device has been dated back to early 19th century, they started to get more commonly used for the past 50 years. Medical devices now are an essential part of healthcare system to diagnose and treat patients [1]. Table 1.1 shows some landmarks in medical device development [1−12].

1.2 What are medical devices

Medical device term is often used for products that have been designed and manufactured for use in healthcare environment. The World Health Organization (WHO) and the US Food and Drug Administration (FDA) define medical device as any instrument, apparatus, machine, appliance, implant, reagent for in vitro use, software, material, intended by the manufacturer to be used, alone or in combination, for human beings, and that does not achieve its primary intended action by pharmacological, immunological, or metabolic means [13,14].

Medical devices are often regulated by different regulatory agencies. For example, medical devices in Europe are regulated by European Medicines Agency (EMA) [15], while in the United Kingdom, medical devices are regulated by Medicines and Healthcare Regulatory Agency [16]. Medical devices in India and China are regulated by their respective agencies, Central Drug Standards Control Organization (CDSCO) [17] and National Medicinal Products Administration (NMPA) [18], respectively. In the United States (US), medical devices are regulated by FDA [14].

Medical devices encompasses broad range of devices (Fig. 1.1) [19−23]: (a) Devices that are used as a final product such as surgical instruments, catheters, coronary stents, vascular grafts, thoracic drains, pacemakers etc. (b) Devices that are used to identify source of the problem by imaging such as magnetic resonance imaging (MRI) machines, X-ray machines etc. (c) Devices that are used as part of the process to make a final product. Examples include MilliporeSigma's Microfiltration and Ultrafiltration cassettes and Ion-Exchange Chromatography Columns that are often used to purify, separate, and retain medicinal molecules such as Insulin, Monoclonal Antibodies (mAbs) and Vaccines.

Table 1.1 Medical device development timeline [1−12].

Year	Medical device(s)
7000 BCE	Scalpels, slings, splints, crutches
1881	Sphygmomanometer
1885	Electrocardiograph machine
1927	Respirator
1933	Electric defibrillators and CPR
1945	Kidney dialysis machine
1948	Plastic contact lens
1951	Artificial heart valve
1952	Pacemaker
1958	Imaging device to detect tumors
1959	Ultrasound
1960	Internal pacemaker
1962	PET transverse section instrument
1969	Portable glucose monitor
1971	Soft contact lens
1978	Cochlear implant
1979	Portable insulin pump
1980−2000	Arthoscope High-resolution imaging devices Cardiovascular systems Ventilators Kidney dialysis machines Neonatal incubators Permanent artificial heart implant Implantable cardioverter defibrillator Computerized axial tomography (CT) scanner Magnetic resonance imaging (MRI) units Coronary stents
2000−2020	Robotics Small precision robotics for orthopedic and neurological surgeries Smart medical capsules Genomics Tissue-engineered products Ultrahigh-field strength MRIs Whole-body CT scanner DEKA prosthetic arm Implantable miniature telescope

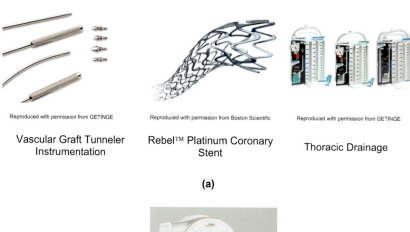

Reproduced with permission from GETINGE
Vascular Graft Tunneler Instrumentation

Reproduced with permission from Boston Scientific
Rebel™ Platinum Coronary Stent

Reproduced with permission from GETINGE
Thoracic Drainage

(a)

Reproduced with permission from Siemens Healthineers

(b)

Magnetom Sola

Figure 1.1 Medical devices used as (a) final product and (b) in identification of the source [20–23].

 ## 1.3 Basic definitions

1. **483**—FDA form that is used by investigators to record observation or practices indicating that an FDA-regulated product may be in violation of FDA's requirements.
2. **513(g)**—Section of the Federal Food, Drug, and Cosmetic Act that provides a means for device manufacturers to obtain information about the Food and Drug Administrator's views regarding the classification of a device.
3. **Contract manufacturer**—Manufacturer that produces a finished device according to the original equipment manufacturers (OEM) specifications.
4. **Corrective and preventive action (CAPA)**—Action taken to eliminate the causes of an existing nonconformity, defect, or other undesirable situation in order to prevent recurrence.

5. **Clinical trial**—Trials that are conducted to determine safety and effectiveness of the product.
6. **Complaint**—Any written, electronic, or oral communication that alleges deficiencies related to the identity, quality, durability, reliability, safety, effectiveness, or performance of a device after it is released for distribution.
7. **Design control**—Set of quality practices and procedures that control the design process to assure that the device meets user needs, intended uses, and specified requirements and can improve and prevent future issues.
8. **Device history file (DHF)**—Compilation of records which describes the design history of a finished device.
9. **Device history record (DHR)**—Records that contain all documentation related to manufacturing and tracking the device and demonstrate that the device is manufactured according to the information in the device master record.
10. **Device master record (DMR)**—Records that contain all of the information and specifications needed to produce a medical device from start to finish, including instructions for all manufacturing processes, drawings, documented specifications, and labeling and packaging requirements
11. **Expanded access**—Also known as "compassionate use," it is a potential pathway for a patient with an immediately life-threatening condition or serious disease to gain access to an investigational medical product for treatment outside of clinical trials when no comparable or satisfactory alternative therapy options are available
12. **Finished device**—Any device or accessory to any device that is suitable for use or capable of functioning, whether or not it is packaged, labeled, or sterilized
13. **Current good manufacturing practices (cGMP)**—FDA mandated regulations that guide the design, monitoring, and maintenance of manufacturing facilities and processes.
14. **Good documentation practices (GDP)**—Standard practice to record raw data entries in a legible, traceable, and reproducible manner.
15. **Investigational device exemption (IDE)**—Practice that allows the investigational device to be used in clinical study in order to collect safety and effectiveness data.

16. **ISO 13485**—An international standard that is designed to be used by organizations involved in the design, production, installation, and servicing of medical devices and related services.
17. **ISO 14971**—An international standard that specifies terminology, principles, and a process for risk management of medical devices, including software as a medical device and in vitro diagnostic medical devices.
18. **ISO 9001**—An international standard that specifies requirements for a quality management system and is used by organizations to demonstrate the ability to consistently provide products and services that meet customer and regulatory requirements.
19. **In vitro diagnostic (IVD)**—Diagnostic tests done on samples such as blood or tissue that have been taken from the human body. The test can detect diseases or other conditions and can be used to monitor a person's overall health to help cure, treat, or prevent diseases.
20. **Medical device reporting (MDR)**—It is a mechanism for FDA and manufacturers to identify and monitor significant adverse events involving medical devices. It defines the device requirements such as description and identification of the device, intended use, labeling, information on the design and manufacture of the device, risk management file, and verification and validation of the device.
21. **Premarket approval (PMA)**—Scientific and regulatory review process adopted by FDA that evaluates the safety and effectiveness of class III medical devices.
22. **Premarket notification (510K)**—It is a premarket submission made to FDA to demonstrate that the device to be marketed is safe and effective, that is, substantially equivalent, to a legally marketed device.
23. **Postmarket surveillance (PMS)**—This is an activity in which medical devices are monitored after they have been cleared for sale and are in use by members of the public to remain complaint with 21CFR Part 820 and ISO 13485.
24. **Quality audit**—It is the process of systematic examination of a quality management system carried out by an internal or external quality auditor or an audit team.
25. **Quality system regulation**—Regulation that requires each manufacturer to establish and maintain a quality system that is appropriate for the specific medical devices designed and manufactured and meets the requirements of FDA 21CFR Part 820.5.

26. **Specification**—Limits between which products or services should operate.
27. **Technical file**—Comprehensive collection of documents that demonstrates products technical basis for conformity to the applicable directives.
28. **Unique device identifier (UDI)**—A system used by the FDA to standardize and identify medical devices through their distribution and use. It is a unique set of alphanumeric codes consisting of both device identifier and a production identifier.
29. **Validation**—Confirmation through provision of objective evidence that requirements for a specific intended use or application have been fulfilled. It often includes the qualification of systems and equipment.
30. **Verification**—Confirmation through provision of objective evidence that specified requirements have been fulfilled and meet design specification.

1.4 Material used in medical devices

Materials that are intended to be used in medical device applications are selected based on their overall benefits to treat patients safely. Early medical device applications used metal such as stainless steel as the primary source of material, but more recent trends have shown that polymers have now surpassed metals due to their inertness to human anatomy, lower cost, and higher effectiveness in patient safety.

The following materials will be discussed in detail:
- Metals
- Polymers

1.4.1 Metals

Metal has been go-to material of choice since the early development of medical devices. They have been used in approximately 80% of all medical devices either by themselves or in combination with other materials. Generally, all metals have good ductility, malleability, mechanical, and electrical properties; however, they are susceptible to corrosion. Medical device industry—preferred metal of choice is stainless steel since it contains iron, carbon, and chromium, which helps prevent oxidation [24,25]. Some examples include tiny screws, catheter reinforcements, and needle cannula to surgical instruments and artificial joints as shown in Fig. 1.2 [26,27].

8 Applications of Polymers and Plastics in Medical Devices

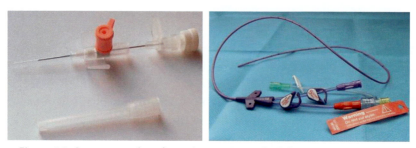

Figure 1.2 Some examples of metal usage in medical device industry [26,27].

In addition to stainless steel, titanium, tantalum, and nitinol have also been widely used. Because of titanium's strength, durability, biocompatibility, corrosion resistance, and flexibility, it is often used for joint replacements, heart valve housings or support, and surgical instruments. Meanwhile, tantalum's joinability and sound dielectric properties make it a preferred candidate for shaped-wire industry. Nitinol is often used in stent applications by medical device manufacturers for its superelasticity and biochemical compatibility [25]. Some of the uses of stainless steel and nitinol in medical devices are shown in Fig. 1.3 [28,29].

Figure 1.3 Uses of stainless steel and nitinol in medical devices [28,29].

Introduction to polymers and plastics for medical devices

Figure 1.4 Magnesium-based bioabsorbable metal screw [30].

Bioabsorbable metals such as magnesium and iron offer additional possibilities for use in medical devices. These metals have similar properties as of bioabsorbable polymers where metals are either absorbed or eliminated by the body after performing their function. However, these metals are susceptible to magnetization due to their inherent material properties causing interference with the equipment. Therefore, these metals cannot be used in implant applications that require MRI. An example of magnesium-based bioabsorbable metal screw is shown in Fig. 1.4 [30].

1.4.2 Polymers

In the past few decades, polymers have made their way into medical device space due to their relatively lower cost, inertness to human anatomy, and superior properties compared with metals. Unlike metals, polymers have poor magnetic properties, which makes them an excellent candidate to be used in scanning machines such as MRI. Polymers are resistant to contamination, can be sterilized, and offer low level of toxicity. In addition, polymer properties can be enhanced by some modifications to make them fit for new applications [24].

Some of the commonly used medical-grade polymers include polyvinyl chloride (PVC), polypropylene (PP), polyethylene (PE), polystyrene (PS), polyamides (PA), polyethylene terephthalate (PET), polyimide (PI), polycarbonate (PC), polyether ether ketone (PEEK), and polyurethane (PU).

Eastman has developed family of copolyesters that are tough, durable, and BPA and halogen-free. Tritan exhibits exceptional clarity, stability, and resistance to heat and chemicals. It also fares well against other polymers and polymer blends such as polycarbonate (PC), PC/ABS, polybutylene terephthalate (PBT), and PET in withstanding hospital disinfection processes.

Typical applications include surgical devices, blood management devices, and renal dialysis devices [31–34]. Like Tritan, Eastman's MXF221 copolyester offers improved chemical resistance while retaining more than 90% of its original impact strength when exposed to strong disinfectants [35]. MXF221 has similar properties like Tritan in durability, stability, and heat resistance [36].

Polyether ether ketone (PEEK) has emerged as a leading thermoplastics material replacing metal implant components in orthopedics [37,38] and trauma [39,40], largely due to their high strength and durability. Some of the applications include plates, nails, and screws for trauma (Piccolo PEEK composite devices from CarboFix: Orthodynamics Ltd.); anchors and interference screws for arthroscopy; finger, hip, and knee components for orthopedics; cranial plates and spinal implants (Lucent® PEEK Interbody Implants from Spinal Elements Inc., Carlsbad California) [41–43].

UHMWPE-based implants are typically used in applications such as orthopedic and cardiovascular implants because of the durability of the material and excellent mechanical strength. UHMWPE examples include use as a graft material in cardiac stents, pacing devices, structural cardiac implants, and total knee arthroplasty (TKA). Fig. 1.5 shows an example of UHMWPE that is used as a replacement material in a total knee replacement [44].

Polyurethanes (PUs) are commonly used material for medical device applications. The segmented structure (hard and soft segment) of PU provides the flexibility and strength. They are tough, biocompatible, hemocompatible, durable, and resistant against chemicals. These properties make them

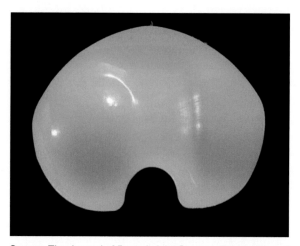

Source: The Journal of Bone & Joint Surgery (British Volume)

Figure 1.5 Ultrahigh-molecular-weight polyethylene (UHMWPE) use in total knee replacement [44].

Introduction to polymers and plastics for medical devices

Figure 1.6 An image of feeding tube made from Polyurethane (PU) [46].

a suitable candidate for medical applications. PU does not contain plasticizers, which makes it a replacement to PVC. PUs are typically used in catheters, feeding tubes, surgical drains, dialysis devices, medical garments, hospital beddings, and more [45]. An example of feeding tube made from Polyurethane is shown in Fig. 1.6 [46].

Polyvinyl chloride (PVC) has been around for decades and is one of the most widely used plastics for medical devices with a market share of approximately 40%. The earliest use of PVC goes back to World War II where it was used as a replacement to glass and metal for packaging of pharmaceutical products [19,47—50].

PVC is a highly versatile polymer, and therefore, it is used in a wide array of products. Some of the properties of PVC include flexibility, antikinking, transparency, ease of fabrication, chemical stability, and biocompatibility. Its antikinking properties allow it to be used for medical tubings. Because it has excellent transparency, it is used in blood bags, dialysis bags, and IV bags. PVC can be both rigid and flexible, and therefore, it is frequently used in rigid and flexible containers. PVC is also one of the materials of choice in production of personal protective equipment (PPE) such as examination and surgical gloves [47—50].

PVC is often the material of choice in hospitals where chemical resistance is needed such as wall coverings. Because of its durability, shock absorbent properties, and noise reduction characteristics, PVC is often used in hospital floors [19,47—50].

1.5 Importance of plastics in medical devices

Introduction of plastic material in medical sphere has completely revolutionized healthcare industry. As our healthcare industry evolves and new technologies are developed, plastic has shown that it is able to adapt with the dynamic nature of the industry, for example, use of plastic in disposable syringes, blood bags, heart valves, intensive care equipment housing, and many other medical devices. Its low cost, durability, and functionality have made plastic useful in prosthetics [24,25,47–50].

Several factors contribute toward determining plastics as a material of choice in healthcare industry including and not only limited to the following:
- Sterilization
- Safety
- Comfort and ease-of-use
- Design flexibility and applications
- Cost
- Other

1.5.1 Sterilization

Importance of sterilization in medical industry is of high importance. Earlier metal-based medical devices were sterilized with an intention to reuse the device due to high cost. However, with the option of single use became more evident, plastics started to be heavily used in medical devices such as surgical gloves, syringes, insulin pens, intravenous tubes, etc. Plastics are also being used to create special antimicrobial touch surfaces that can repeal bacteria and other microbes, thus reducing the spread of dangerous diseases. Additionally, based on the application, more and more medical devices are now required to be sterilized and cleaned prior to packaging.

1.5.2 Safety

Plastics are known to be a durable material and have been widely used in toy industry. Some examples include child proof door lock, plastic plugs for power sockets, etc. Similarly, the durability nature of plastics allows them to be used in applications where safety is of paramount importance such as tamper-proof caps on medical packaging, blister packs, and biohazard waste disposal bags. The biohazard waste disposal bags are used for transporting medical wastes. Use of plastics in face shields has been important in protecting medical professionals from coronavirus and preventing them from getting exposed to infectious diseases.

1.5.3 Comfort and ease-of-use

With the development of new technologies and new emerging materials, the healthcare industry is moving away from metals that were historically used to treat patients. For example, metals were used to treat patients in need of prosthetics. With the plastic emerging as the material of choice based on its physical, mechanical, and chemical characteristics, it is now used as a replacement to such metal components. Plastics not only provide improved comfort but also provide specialized solutions that allow for healthier and normal lives.

1.5.4 Design flexibility and applications

Design flexibility is another factor that differentiates plastics from metals. Processing plastic into final product requires melting polymer resin into a mold, which takes the shape of final product. Plastics can be processed into endless shapes and products per the requirement of specific applications. Some examples of the wide array of plastics-based medical devices include tubing, syringes, intravenous bags, catheters, labware, films for packaging, surgical instruments, pacemakers, housings and connectors, stents, or joint replacement devices.

1.5.5 Cost

In comparison with metals, plastics can not only be mass-produced but at cost-effective rates, and for a wider range of applications. Plastics are known to be tough which can withstand large amount of stress. Unlike metals, they are not susceptible to corrosion due to their inertness. All these properties relate to overall cost, as it minimizes the need for secondary processes, which in turn minimizes the time and effort that would otherwise go into the upkeep of said metal devices.

1.6 General requirements of materials to qualify for medical devices

To identify and qualify a material, a thorough understanding of polymer morphology and structure, additives, and its impact on properties is required. There are three key steps that need to be taken to select a material for medical applications [51]:
- Preliminary assessment
- Screening
- Ranking

To select appropriate candidate material, a thorough defined application requirements need to be identified. List can be narrowed down to two or three candidates after careful consideration, and then the final selection should be determined by testing.

1.6.1 Step 1—preliminary assessment

The first step of the process is to perform a preliminary assessment on a range of polymers to determine if they meet the need of the final product. This assessment is largely done to identify polymers whether they meet certain defined requirements. Based on the outcome, a decision is made to accept or reject a given polymer. The following is the list of requirements that each polymer needs to satisfy to move to step 2:

(a) Environmental
- Biocompatibility
- Exposure to active ingredient such as drug
- Type of use—single-use or multiuse
- Sterilization
- Chemical resistance
- Secondary operations required
- Exposure to humidity and temperatures
- Dimensional stability
- UV resistance
- Flame retardancy
- Color
- Application

(b) Functional
- Part dimension and dimensional stability
- Load on the part, duration of exposure
- Maximum stress
- Mechanical properties
- Wear resistance
- Electrical insulation properties
- Special processing techniques
- Target product cost
- Projected life of the part or design
- Final assembly requirement

1.6.2 Step 2—screening

Step 2 of the process is to screen polymers that have met both the environmental and mechanical/physical requirements on the preliminary assessment in step 1. Each polymer is further screened against defined requirements mentioned in step 1. The list of polymers is narrowed down to a maximum of three and ranked based on overall match to the identified requirements.

1.6.3 Step 3—rank

The third step of the process is to rank each polymer based on the plastics material index on objectives that define the performance of the product. Based on the resulting rank, best choice of polymer is chosen.

1.7 Global nature of medical devices

The benefit of using medical devices has helped healthcare providers to diagnose, treat, and assist patients in improving their quality of life. During the 7-year period (2018–2025), it is projected that there will be a surge in the global device market due to the increase in population with chronic conditions as well as growth in selective surgeries [52,53].

The global medical device market is segmented by the type of equipment and the area they serve. List includes in vitro diagnostics (IVD), dental equipment and supplies, ophthalmic devices, diagnostic imaging equipment, cardiovascular devices, hospital supplies, surgical equipment, orthopedic devices, patient monitoring devices, diabetes care devices, nephrology and urology devices, ENT devices, anesthesia and respiratory devices, neurology devices, and wound care devices [50]. In 2018, the global medical device market was valued at $423.8 billion. The market grew at a compound annual growth rate (CAGR) of 5.28% since 2014 and is expected to grow at a CAGR of 5.33% to nearly $521.64 billion by 2022 [53].

The largest segment was the cardiovascular devices market which saw growth at a CAGR of 15.2% of the total in 2018 to nearly 64.5 billion. It was followed by IVD, diagnostic imaging equipment, and then other segments. It is projected that the fastest growing segments in the medical devices market will be the patient monitoring devices and diabetes care devices, which expects to grow at CAGRs of 8.2% and 7.6%, respectively. Hospitals and clinics were the largest segments of the medical devices market, which saw growth rise to $372.59 billion in 2018 [53].

North America had the largest global medical device market share in 2018, a 42.6% of the total market. It was followed by Western Europe, Asia—Pacific, and then the other regions. It is projected that the fastest growing regions in the medical devices market will be Asia—Pacific and South America, where growth will be at CAGRs of 7.07% and 6.53%, respectively. These will be followed by Africa and the Middle East where the markets are expected to grow at CAGRs of 6.47% and 6.43%, respectively [50]. It is projected that the cardiovascular device segment will see significant growth opportunities with gain of $16.17 billion of global annual sales by 2022. The United States will see significant increase in the medical devices market size topping at $35.33 billion [53].

1.8 Socioeconomic factors

Socioeconomic factors play an important role in defining the healthcare system. The economic disparity in the United States compounded together with increased cost of medical care has impacted an average American with one in three deaths annually accounting to social factors. It is important for healthcare industry to understand these factors, become proactively involved, and take actions that will improve overall patient care experience. Some socioeconomic factors are discussed in the following [54]:
1. Graying populations require more medical care
2. Cost of care
3. Proliferation of procedures that require more medical care
4. Device design to reduce skill requirements for use
5. Role of polymers in cost reduction

1.8.1 Graying populations require more medical care

The WHO estimates in its Global Health and Aging report that people with age 65 or older are projected to grow from 524 million to 1.5 billion over a period of 40 years, with majority of increase happening in developing countries. This increase in life expectancy may be as a result of change in the leading cause of death from infections to chronic noncommunicable diseases such as hypertension, high cholesterol, arthritis, diabetes, heart disease, cancer, dementia, and congestive heart failure [55,56].

As the life expectancy of people aged 65 years or older increases, it opens a bigger question of how this impacts the healthcare system. The Office of Disease Prevention and Health Promotion projects that by 2030, the number of chronic conditions along with the level of disability will put strain and increase demand on the healthcare system and, consequently, put financial burden on the patient [57,58].

Reducing medical cost is a two-step process, which involves identifying and overcoming challenges and then implements a strategic plan that focuses on patient care [58–60].

There are several challenges that the healthcare system needs to overcome:
(a) Resource needs that will continue to rise
(b) Increase in obesity within society
(c) Shortage of healthcare professionals
(d) Ratio of diversity of caregivers to diversity of patients
(e) Focus of care on single disease control
(f) Impact of family structure on overall care

Implementation of strategic plan is a second part of a two-part process. Within this plan, the healthcare system should assume that there will be an increase in chronic conditions among aging population and figure out ways to implement new approaches that will deliver the healthcare. Another aspect of healthcare is to focus on specific illness rather than to focus on two or more conditions, which could result in inefficient attention to the patient [59]. The healthcare system should also implement multidisciplinary approach so that patient receives better care. They should take a proactive approach by providing preventive care rather than reactive care [58].

1.8.2 Cost of care

The cost of care has been steadily rising over the past decade impacting both insured and uninsured people who are in worst health condition being the most affected. The National Health Interview Survey did a survey among adults and found out that approximately 9% or 1 in 10 adults delayed or did not receive medical care because of its cost [61].

Nearly one in five adults in worse health ($\sim$19%) delayed or did not receive medical care due to cost barriers compared with 7% adults in better health [61].

According to the Kaiser Family, foundation analysis of the National Health Interview Survey showed that approximately 28 percent uninsured adults or one in four delayed or went without healthcare because of cost reasons. Meanwhile, 7% of adults with health insurance encountered cost-related access barriers to care [61].

During the period between 1998 and 2007, uninsured adults have consistently experienced more difficulty accessing healthcare due to cost [61].

1.8.3 Proliferation of procedures that require more medical care

With increasing aging population, the healthcare in the United States is seeing a rapid increase in the cost of care [58,59]. On average, per capita healthcare spending in the United States has reached 50%—200% of other economically developed countries [62—64]. In contrary, the United States is trailing the world economies in life expectancy and quality of care [64,65].

The disparity of care is tied to the level of quality as reported by Dartmouth Atlas of Healthcare. Dartmouth Institute for Health Policy did a research where they examined patients hospitalized between 1993 and 1995 for hip fracture, colorectal cancer, or acute myocardial infarction and compared them to Medicare enrollees in their past 6 years of life and measured how much each member spent on end-of-life care [66]. Researchers found out that healthcare spending and quality varied based on the geographical location within the United States. Area that had more medical practitioners per capita saw increased healthcare spending among examined patients. On the other hand, medicare beneficiaries who lived in high-spending areas received approximately 60% more services than did those who lived in low-spending areas [62,66].

In another study, researchers measured the relationship between intensity of care and quality in New York State by randomly assessing patients that were sent to hospitals through ambulance. They found that high-intensity care may, infact, improve patient outcome [67].

Silber et al. tried to understand relationship between care intensity and mortality among Medicare beneficiaries between 2000 and 2005. They found that greater care intensity resulted in lower mortality during the first 30 days after admission. They found that mortality was no longer dependent on patients care from lower- and higher-intensity hospitals after the first 30-day window post hospital admission [68,69].

1.8.4 Device design to reduce skill requirements for use

Device design is a key factor in determining skill requirements for use. Interaction between user (device operator) and device usability is termed as human factors/usability engineering. Fig. 1.7 presents a three-layered model

Introduction to polymers and plastics for medical devices

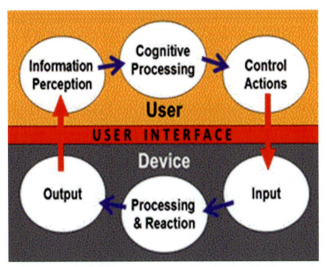

Figure 1.7 Device user interface in operational context [70]. *Source: Food & Drug Administration, Republished on December 14th, 2021.*

that shows interactions between a user and a device, the processes performed by each, and the user interface between them. The device user interface, depicted as the red zone, is the critical element in user—device interactions and is used to design the user interface. User interface includes all components with which users interact while preparing the device for use, using the device, or performing maintenance [70].

To understand complete functionality of user-device system, it is imperative to understand how each system work individually:

1.8.4.1 User system
1. Information perception—perceive information from the device.
2. Cognitive processing—interpret the information and make decisions about next steps.
3. Control actions—manipulate the device, its components, and/or its controls.

1.8.4.2 Device system
1. Receives input from the user.
2. Processing and reaction.
3. Provides output to the user.

Human factor/usability engineering is important to medical devices, as it minimizes use-related hazards and risks and then confirms that those efforts were successful, and users can use the device safely and effectively. Some of the benefits include the following [70]:
- Easy to use
- Provide safer connections between device components and accessories
- Easier-to-read controls and displays
- Better user understanding of the device's status and operation
- Better user understanding of a patient's current medical condition
- More effective alarm signals
- Easier device maintenance and repair
- Reduced risk of use error
- Reduced risk of adverse events
- Reduced risk of product recalls

1.8.5 Role of polymers in cost reduction and increasing the availability of devices

Polymers are the material of choice in medical devices because of their added benefits over metal. Some examples include drug delivery devices, implantable medical devices, biodegradable implants, active pharmaceutical ingredients, medical-grade coatings, and many more [71].

The cost of medical device is driven by factors, which play an important role. These factors are as follows:
- Material cost or processability
- Product weight
- Materials of construction
- New technologies

1.8.5.1 *Material costs or processability*

Cost of material and the final device cost determine the viability of the product. Large medical devices are price sensitive, and any effort to reduce the final product cost is always considered. For example, by selecting a material with higher temperature properties, the time it takes to sterilize the product can be significantly reduced. This is beneficial because the time it takes to sterilize and ultimately manufacture the product would be significantly reduced leading to cost savings [72]. Material properties are tied-in with processability. Any material that has the potential to reduce the cost of manufacturing can bring added benefit to the device manufacturer. For example, any changes to processing parameter such as injection rate or moldability that improves processing and reduces scrap would be considered highly valuable [72].

1.8.5.2 Product weight
Weight of the product could bring in some added cost to the manufacturer primarily from increased transportation costs and material standpoint. To mitigate, some manufacturers are shifting toward using raw materials that are less dense but bring in same or better properties lowering not only the cost of the material but also the transportation costs. One example includes the surgical instruments that, by reducing the overall weight of the product, it gives a functional advantage allowing greater ease of use and resulting in less fatigue in arms and hands of the surgeon during surgery. Another aspect of lowering the weight of the product is to lower the cost of disposal of the product at end of its life cycle. In addition, smaller footprint requires lesser packaging that directly results in lower shipping and handling costs without the loss of functionality and quality of the product [72].

1.8.5.3 Materials of construction
There are several advantages of using a simplified approach where a common polymer is used to produce both components and devices. Although this may not be common, the use of common material helps drive material costs down, and it is also an opportunity for a manufacturer to expand the products. It is also beneficial for the manufacturer; as the consumption of a polymer grows, there is an opportunity for a better purchasing price. Another advantage of using a single material will be not only to have a lesser complex qualification process but also the need to qualify the same material across multiple product lines. There is no need to stock multiple grades or different types of plastics, and thereby eliminating the need to perform incoming testing on all the materials [72]. Although it is recommended to identify a single material source for components and devices, an additional consideration needs to be taken to build the product on the one material that works. For example, the material should offer properties that are required in a product, be able to withstand exposure to sterilization methods, use similar processing approach, do not require special technology requirements, etc. This collective understanding generates more efficient manufacturing operations, a better product, and a lower product cost.

1.8.5.4 New technologies

Technology plays a crucial role in developing new products as well as defining overall cost of the product. Inefficient or older technologies can result in loss of product and time, which further results in higher product cost. Although new technologies require initial capital investment, it can lead in minimizing inefficiencies, improving manufacturability of the product, and further reducing product cost. For example, molding a device using a cold runner technology can result in higher scrap that is either discarded or reused. This costs a lot of money and time. On the other hand, new technology can improve products and result in better market share and reduced product costs. An example for this kind of situation may be use of hot runner systems that eliminates total loss of material.

Another example includes vision systems and robotics that are installed on manufacturing line. Vision systems can measure tight tolerance parts, while robotics move product from one place to another, thereby eliminating the need for human handling. New precision technology such as micromolding makes it possible to produce microsized parts, which allows less material to be consumed and ultimately results in lower product cost. By implementing new technologies, manufacturers are providing customers with a defect-free product with high quality and lower product cost [72].

References

[1] Medical Devices, Managing the Mismatch: An Outcome of the Priority Medical Devices Project, World Health Organization, Geneva, 2010. https://apps.who.int/iris/bitstream/handle/10665/44407/9789241564045_eng.pdf?sequence=1.
[2] L.R. Atles, A Practicum for Biomedical Engineering and Technology Management Issues. Dubuque, Kendall Hunt Publishing, 2008.
[3] Medical Device Timeline, Morgridge Institute for Research, Madison WI. https://morgridge.org/outreach/teaching-resources/medical-devices/medical-devices-timeline/.
[4] J. Gaev, Technology in health care, in: J. Dyro (Ed.), Clinical Engineering Handbook, Elsevier Academic Press, Burlington, 2004, pp. 342–345.
[5] M. Butter, et al., Robotics for Healthcare, Final Report, European Commission, DG Information Society, 2008.
[6] W. Kaplan, et al., Future Public Health Needs: Commonalities and Differences between High- and Low-Resource Settings, World Health Organization, Geneva, 2010.
[7] Special report on health care technology, Economist (2009).
[8] R.E. Geertsma, et al., New and Emerging Medical Technologies: A Horizon Scan of Opportunities and Risks, National Institute for Public Health and the Environment (RIVM), 2007. Report 360020002.
[9] Computer-Aided Surgery: a GPS for the OR, Health Devices 38 (7) (2009) 206–218.
[10] Top 10 Hospital Technology Issues: C-Suite Watch List for 2009 and Beyond, Emergency Care Research Institute, 2009.

[11] Robot-assisted Radical Prostatectomy: A Technology Assessment, California Technology Assessment Forum, 2008.
[12] Radiation-emitting Products and Procedures Medical Imaging, Computed Tomography (CT), United States Food and Drug Administration, Washington DC, 2002.
[13] Medical Devices: Full Definitions. World Health Organization. Geneva. https://www.who.int/medical_devices/full_deffinition/en/.
[14] Medical Devices Overview, United States Food and Drug Administration, Washington DC, 2018. https://www.fda.gov/industry/regulated-products/medical-device-overview#What%20is%20a%20medical%20device.
[15] Medical Devices, European Medicines Agency (EMA). https://www.ema.europa.eu/en/human-regulatory/overview/medical-devices.
[16] Organizations, Medicines & Healthcare Products Regulatory Agency (MHRA), UK. https://www.gov.uk/government/organisations/medicines-and-healthcare-products-regulatory-agency.
[17] Directorate General of Health Services, Central Drugs Standard Control Organization (CDSCO), India. https://cdsco.gov.in/opencms/opencms/en/Home/.
[18] National Medical Products Administration (NMPA), China. http://english.nmpa.gov.cn/.
[19] V.R. Sastri, Plastics in Medical Devices: Properties, Requirements and Applications, Elsevier, 2010.
[20] Vascular Graft Tunneler Instrumentation, GETINGE, Merrimack NH. https://www.getinge.com/int/product-catalog/vascular-tunneler/.
[21] REBEL™ Platinum Chromium Coronary Stent System, Boston Scientific, Natick MA. https://www.bostonscientific.com/en-US/products/stents–coronary/rebel-platinum-chromium-coronary-stent-system.html.
[22] Thoracic Drainage, GETINGE, Merrimack NH. https://www.getinge.com/us/solutions/acute-care-therapies/thoracic-drainage/.
[23] Magnetom Sola, Siemens Healthineers, Erlangen Germany. https://www.siemens-healthineers.com/en-us/magnetic-resonance-imaging/0-35-to-1-5t-mri-scanner.
[24] Materials Science in Medical Device Manufacturing, Proven Process Medical Devices, Mansfield MA. https://provenprocess.com/medical-device-engineering/materials-science.
[25] M. Barbella, Practical Matters: A Look at Medical Device Materials, Medical Product Outsourcing, 2018. https://www.mpo-mag.com/issues/2018-03-01/view_features/practical-matters-a-look-at-medical-device-materials/.
[26] Two-Lumen Polyurethane Catheter, Wikimedia Commons, Free Media Repository, July 2nd, 2014. https://commons.wikimedia.org/wiki/File:PICC_line.jpg.
[27] Venflon Intravenous Cannula, Wikimedia Commons, Free Media Repository, April 12th, 2006. https://commons.wikimedia.org/wiki/File:Venflon_Intravenous_Cannula_2.jpeg.
[28] Titanium Dental Implants, Wikimedia Commons, Free Media Repository, February 2nd, 2011. https://commons.wikimedia.org/wiki/File:Teeth.jpg.
[29] T. Duerig, A. Pelton, D. Stockel, An overview of nitinol medical applications, Mater. Sci. Eng. 273–275 (1999) 149–160.
[30] R. Biber, J. Pauser, M. Geßlein, H.J. Bail, Magnesium-Based Absorbable Metal Screws for Intra-Articular Fracture Fixation, in: Case Reports in Orthopedics. Germany, Hindawi Publishing Corporation, London, 2016, pp. 1–4.
[31] NuGen™ Eastman Tritan Copolyester Syringes. DMC Medical. http://www.dmcmedical.net/medical-products/nugen-eastman-tritan-copolyester-syringes/.
[32] Medical, Eastman, Kingsport TN. https://www.tritanmoldit.com/medical.
[33] Disinfect with Confidence, Eastman Literature Center, Eastman, Kingsport TN. https://www.eastman.com/Literature_Center/S/SPMBS5125.pdf.

[34] A Preferred Supplier that Goes the Extra Mile, Eastman Regulatory and Technical Services, Eastman, Kingsport TN. https://www.tritanmoldit.com/blog-categories/processing?page=1.
[35] Eastman Tritan™ Copolyester, Eastman, Kingsport TN. https://www.eastman.com/Markets/Medical-Devices/Blood-Contact/Pages/Overview.aspx.
[36] Eastman MXF221 Copolyester, Eastman, Kingsport TN. https://www.eastman.com/Pages/ProductHome.aspx?product=71110599.
[37] K. Liao, Performance characterization and modeling of a composite hip prosthesis, Exp. Tech. (1994) 33−38.
[38] G.R. Maharaj, R.D. Jamison, Intraoperative impact: characterization and laboratory simulation on composite hip prostheses, in: R.D. Jamison, L.N. Gilbertson (Eds.), STP 1178: Composite Materials for Implant Applications in the Human Body: Characterization and Testing, ASTM, Philadelphia, 1993, pp. 98−108.
[39] D.J. Kelsey, G.S. Springer, S.B. Goodman, Composite implant for bone replacement, J. Compos. Mater. 31 (16) (1997) 1593−1632.
[40] A.A. Corvelli, P.J. Biermann, J.C. Roberts, Design, analysis, and fabrication of a composite segmental bone replacement implant, J. Adv. Mater. (1997) 2−8.
[41] S. Lovald, S.M. Kurtz, Applications of Polyetheretherketone in Trauma, Arthroscopy, and Cranial Defect Repair. William Andrew, Applied Science Publishers, 2012, pp. 243−260.
[42] PEEK Interference Screws, Tulpar Medical Solutions, Yenimahalle-Ankara, Turkey. https://tulparmed.com/peek-interference-screws/.
[43] Spinal Elements Launches PEEK Interbody Implant with Titanium Porous Coating, Spinal Element, Carlsbad CA. https://orthostreams.com/2012/08/spinal-elements-launches-peek-interbody-implant-with-titanium-porous-coating/.
[44] G. Blunn, E.M. Brach del Preva, L. Costa, J. Fisher, M.A.R. Freeman, Ultra-high molecular weight polyethylene (UHMWPE) in total knee replacement: fabrication, sterilization and wear, J. Bone Joint Surg. 84-B (7) (2002) 946−949.
[45] S.E. Westerdale, Thermoplastic Polyurethane for Healthcare Applications, Medical Design Briefs, 2013.
[46] Enteral Feeding Tube Stylet Retracted, Wikimedia Commons, June 7th, 2018. https://commons.wikimedia.org/wiki/File:Enteral_Feeding_Tube_Stylet_Retracted.png.
[47] T. Johnsen. When Plastics Revolutionized Healthcare − Medical Devices in a Historical Perspective, PVC Med Alliance. Brussels, Belgium. https://pvcmed.org/healthcare/when-plastics-revolutionised-healthcare/.
[48] C.R. Blass, Polymers in Disposable Medical Devices: A European Perspective, iSmithers Rapra Publishing, 1999.
[49] C.R. Blass, The Role of Poly(Vinyl Chloride) in Healthcare, iSmithers Rapra Publishing, 2001.
[50] X. Zhao, J.M. Courtney, Update on Medical Plasticized PVC, iSmithers Rapra Publishing, 2009.
[51] An Overview of the Plastic Material Selection Process for Medical Devices, HCL Technologies, Noida, India, 2013, in: https://www.hcltech.com/white-papers/engineering-rd-services/overview-plastic-material-selection-process-medical-devices.
[52] Key Market Insights - Medical Device/Medical Devices Market. Report ID. FBI100085. Fortune Business Insights. https://www.fortunebusinessinsights.com/industry-reports/medical-devices-market-100085.
[53] Global Medical Devices Market Report 2019-2022 − A $521+ Billion Opportunity Analysis, Research and Markets, Dublin Ireland, 2019. https://www.globenewswire.com/news-release/2019/09/19/1918062/0/en/Global-Medical-Devices-Market-Report-2019-2022-A-521-Billion-Opportunity-Analysis.html.

[54] Socioeconomic Health Attributes, LexisNexis Risk Solutions. https://risk.lexisnexis.com/products/socioeconomic-health-attributes.
[55] A. Garza, The Aging Population: The Increasing Effects on Healthcare, Pharmacy Times, 2016. https://www.pharmacytimes.com/publications/issue/2016/january 2016/the-aging-population-the-increasing-effects-on-health-care.
[56] R. Suzman, J. Beard, Global Health and Aging: Preface, National Institute on Aging, 2011. Updated January 22, 2015, www.nia.nih.gov/research/publication/global-health-and-aging/preface (Accessed 1 August 2015).
[57] Older Adults. HealthyPeople.gov, www.healthypeople.gov/2020/topics-objectives/topic/older-adults. (Accessed 1 August 2015).
[58] When I'm 64: How Boomers Will Change Health Care, American Hospital Association., Chicago, IL, 2007. www.aha.org/systems/files/content/00-10/070508-boomerreport.pdf.
[59] L.P. Fried, D.J. Ferrucci, J.D. Williamson, G. Anderson, Untangling the concepts of disability, frailty, and comorbidity: implications for improved targeting and care, J. Gerontol. A Biol. Sci. Med. Sci. 59 (3) (2004) 255−263.
[60] Population aging will have long-term implications for economy, Natl. Acad. Sci. (2012). www.sciencedaily.com/releases/2012/09/120925143920.htm (Accessed 1 August 2015).
[61] G. Claxton, B. Sawyer, C. Cox, How Does Cost Affect Access to Care? Peterson-KFF Health System Tracker, 2019. https://www.healthsystemtracker.org/chart-collection/cost-affect-access-care/#item-adults-who-are-in-worse-health-have-more-difficulty-accessing-care-due-to-cost_2017.
[62] L.A. Burke, A.M. Ryan, The complex relationship between cost and quality in US Healthcare, Virtual Mentor, AMA J. Ethics 16 (2) (2014) 124−130, https://journalofethics.ama-assn.org/article/complex-relationship-between-cost-and-quality-us-health-care/2014-02.
[63] G.A. Cuckler, A.M. Sisko, S.P. Keehan, et al., National health expenditure projections, 2012-22: slow growth until coverage expands and economy improves, Health Aff. 32 (10) (2013) 1820−1831.
[64] Health at a Glance 2013: OECD Indicators. Organization for Economic Cooperation and Development. OECD Health Data 2013: How Does the United States Compare, 2013. http://www.oecd.org/unitedstates/Briefing-Note-USA-2013.pdf.
[65] K. Davis, C. Schoen, K. Stremikis, Mirror, Mirror on the Wall: How the Performance of the US Health Care System Compares Internationally: 2010 Update, The Commonwealth Fund, 2013. https://www.commonwealthfund.org/publications/fund-reports/2010/jun/mirror-mirror-wall-how-performance-us-health-care-system.
[66] E.S. Fisher, D.E. Wennberg, T.A. Stukel, D.J. Gottlieb, F.L. Lucas, E.L. Pinder, The implications of regional variations in Medicare spending. Part 1: the content, quality, and accessibility of care, Ann. Intern. Med. 138 (4) (2003) 273−287.
[67] J.J. Doyle, J.A. Graves, J.A. Gruber, S. Kleiner, Do High-Cost Hospitals Deliver Better Care? Evidence From Ambulance Referral Patterns, National Bureau of Economic Research, 2013. http://www.nber.org/papers/w17936.pdf?new_window=1.
[68] J.H. Silber, R. Kaestner, O. Even-Shoshan, Y. Wang, L.J. Bressler, Aggressive treatment style and surgical outcomes, Health Serv. Res. 45 (6 Pt 2) (2010) 1872−1892.
[69] A.E. Barnato, Comment on Silber et al. Investing in postadmission survival—a "failure-to-rescue" US population health, Health Serv. Res. 45 (6 Pt 2) (2010) 1903−1907.

[70] Human Factors and Medical Devices, United States Food and Drug Administration (FDA), Silver Spring MD. https://www.fda.gov/medical-devices/device-advice-comprehensive-regulatory-assistance/human-factors-and-medical-devices.
[71] E. Price, Top 4 Costs Associated With Medical Grade Polymer Synthesis, SEQENS North America, 2017. https://www.pcisynthesis.com/top-4-costs-associated-with-medical-grade-polymer-synthesis/.
[72] L. Czuba, Applications of plastics in medical devices and equipment, in: Handbook of Polymer Applications in Medicine and Medical Applications, 2014, pp. 9–19. https://www.ncbi.nlm.nih.gov/pmc/articles/PMC7151894/.

CHAPTER TWO

Classification of medical devices

Contents

2.1 Role of Food and Drug Administration	27
2.2 Classification of medical devices	28
2.2.1 Class I devices	28
2.2.2 Class II devices	31
2.2.3 Class III devices	31
2.3 Why are devices classified?	33
2.4 Device classification panels	33
2.5 Exemptions	33
2.6 Emergency use authorization	35
2.7 Premarket notification, 510(k)	36
2.8 Premarket approval	36
2.9 Postapproval market requirements	38
2.9.1 General requirements	38
2.9.2 Postapproval report	39
2.9.3 Premarket supplement	39
2.9.4 Medical Device Reporting	40
2.9.5 Postmarket surveillance studies	40
2.10 FDA's Accelerated Approval Program	40
2.11 FDA's Breakthrough Devices Program	41
2.11.1 Benefits	41
2.11.2 Program principles	41
2.11.3 Program features	42
References	42

2.1 Role of Food and Drug Administration

The US Food and Drug Administration (FDA) is a federal agency in the United States that regulates products critical to health and safety of the public. Some of its roles and responsibilities include [1—4]:
- Development of human and veterinary biological products
- Development of medical devices
- Cosmetics and devices that emit radiation
- Manufacturing, marketing, and distribution of tobacco products
- Participating in the development and use of standards

Enforcing laws regulate safety and effectiveness of all drugs, biologics, and medical devices.

2.2 Classification of medical devices

The FDA regulates all medical devices marketed in the United States. Medical devices are classified based on their risks and regulatory controls required for adequate safety and performance. According to the FDA, medical devices are classified into three categories based on the degree of risk they pose to the patient and/or user [5–7]:
- Class I
- Class II
- Class III

Class I devices are the ones that have the lowest risk to the patient, whereas class III devices can cause potential harm to the patient. Device classification also depends on its intended use and upon indications for use. Figure 2.1 shows some examples [8–11].

The FDA has established classifications for approximately 1700 different generic types of devices and grouped them into 16 medical specialties referred to as panels. Based on the level of control needed to assure the safety and effectiveness of each device, the FDA has assigned each medical device into one of the three regulatory classes as shown in Table 2.1 [6,7].

The class to which your device is assigned determines the type of premarketing submission/application required for FDA clearance to market. A 510(k), premarket notification, is required for marketing if the device is classified as class I or II and not exempted. All devices classified as exempt are subject to the limitations on exemptions, which are covered under 21 CFR Parts 862–892. For devices that are classified as class III, a premarket approval (PMA) is required [6,7].

2.2.1 Class I devices

The FDA defines class I devices as devices which are:

> not intended for use in supporting or sustaining life or of substantial importance in preventing impairment to human health, and they may not present a potential unreasonable risk of illness or injury

Class I devices are the most common class of devices regulated by the FDA. It comprises approximately 47% of approved devices on the market. These devices do not come in contact with patient's internal organs and

Classification of medical devices

Class I Devices

Class II Devices

Reproduced with permission from Boston Scientific

Class III Devices

Figure 2.1 Some examples of class I, II, and III medical devices [8–11].

Table 2.1 Regulatory classes assigned to each medical device [6,7].

Class I general control	Class II general control	Class III general controls
• With exemptions	and special controls	and premarket
• Without exemptions	• With exemptions	approval
	• Without exemptions	

cardiovascular system and have low impact on patient's overall health. Although majority of class I devices are exempt from premarket notification and PMA and are easier and faster to bring to market, they are not exempt from FDA general controls that addresses adulteration, misbranding, device registration, records, and good manufacturing practices. These devices are subject to the fewest regulatory requirements [6,7]. Some examples of class I devices are shown in Fig. 2.2 [12−16].

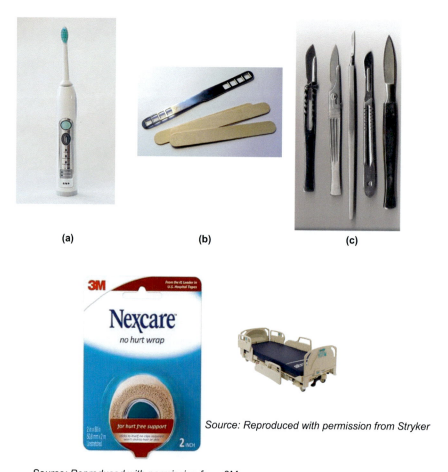

Figure 2.2 (a) Electric toothbrush, (b) tongue depressor, (c) reusable surgical scalpel, (d) 3M Nexcare Hurt Wrap, and (e) Stryker's hospital bed [12−16].

2.2.2 Class II devices
The FDA defines class II devices as follows:

devices for which general controls are insufficient to provide reasonable assurance of the safety and effectiveness of the device

Class II medical devices present a higher risk than class I devices due to their complex nature and potential contact with patient's internal organs and cardiovascular system. In addition to being subjected to FDA's general controls, these devices are also subjected to special controls because general controls are insufficient to provide reasonable assurance for the safety and effectiveness of the device.

Majority of class II devices are approved for market through premarket notification, 510(k), and PMA process. By using this process, manufacturers can demonstrate that a device is safe and effective [6,7]. As part of this process, an assessment requires comparison of the new device with a predicate device.

A predicate device is a medical device that is used as a point of comparison for approval. Medical device manufacturers often utilize 510(k) de novo submission process for a lower risk device without a predicate device. By using this process, the manufacturer proposes list of special controls that provide reasonable assurance of safety and effectiveness of the device instead of providing evidence of equivalence to a predicate device [17].

Some examples of class II devices as shown in Fig. 2.3 include [18—22]:

2.2.3 Class III devices
The FDA defines class III devices as follows:

devices which usually sustain or support life, are implanted or present a potential unreasonable risk of illness or injury.

Class III devices present the highest risk to the patient and/or user since they come directly in contact with the patient and may cause severe harm. These devices are highly regulated by the FDA and comprise approximately 10% of approved devices on the market. Some class II devices may be designated as class III if the manufacturer is unable to demonstrate substantial equivalency to an existing product during 510(k) filing process [6,7]. Some examples of class III medical devices include breast implants, pacemakers, defibrillators, ventilators, cochlear implants, and implanted prosthetics [23—26] (Fig. 2.4)

32 Applications of Polymers and Plastics in Medical Devices

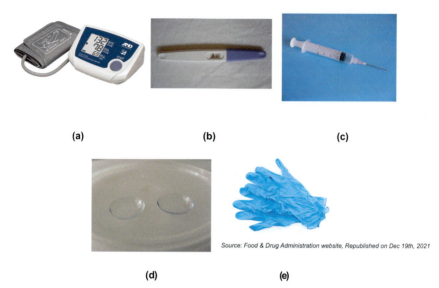

(a) (b) (c)

(d) (e)

Figure 2.3 (a) Blood pressure cuffs, (b) pregnancy test kit, (c) syringes, (d) contact lens, and (e) surgical gloves [18–22].

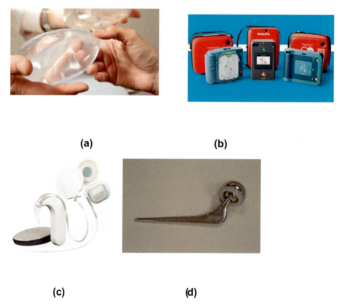

(a) (b)

(c) (d)

Figure 2.4 (a) Breast implants, (b) defibrillator, (c) cochlear implants, and (d) implanted prosthetics [23–26]. *Source: Food & Drug Administration website, Republished on Dec 19th, 2021*

In addition to being subjected to FDA's general and special controls, these devices have to go through FDA's PMA process. This is primarily because general and special controls are not sufficient to assure the safety and effectiveness of class III devices [6,7].

2.3 Why are devices classified?

To bring medical device to market, it is important to determine the correct classification of the medical device. The correct classification will determine the regulatory requirements a device would require bringing it to market. It is important for any device maker to be compliant during product development through commercialization. This will not only ensure compliance toward FDA Code of Federal Regulations (CFR) but also avoid additional audits and quicker medical device approval by the FDA [27].

2.4 Device classification panels

Device classification panels are list of medical speciality areas with regulation citation that are associated with each medical device. The FDA has organized 1700 medical devices in Title 21 of the Code of Federal Regulations (CFR) into 16 medical specialty panels, for example, cardiovascular devices or dental devices (Table 2.2) [28]. For each medical device, CFR provides a general description, its intended use, the class to which it belongs, and marketing requirements.

2.5 Exemptions

When a medical device is not required to provide reasonable assurance of safety and effectiveness for the device, it is considered as exempted. Certain class I and class II devices are exempt from 510(k) requirements and quality system regulations [29].

Table 2.3 shows list of class I and class II medical devices [30]. Majority of class I devices are exempted by FDA; however, it is important to confirm the exempt status and any limitations that apply with 21 CFR Parts 862–892. If a device falls into a generic category of exempted class I devices, a premarket notification application, 510(k), and FDA clearance are not required for launching the device in the United States [30].

The FDA implemented Modernization Act of 1997 (FDAMA) that would redirect resources from reviewing premarket notification submission

Table 2.2 Device classification panels [28].

Medical specialty	Regulation citation (21CFR)
Anesthesiology	Part 868
Cardiovascular	Part 870
Chemistry	Part 862
Dental	Part 872
Ear, nose, and throat	Part 874
Gastroenterology and urology	Part 876
General and plastic surgery	Part 878
General hospital	Part 880
Hematology	Part 864
Immunology	Part 866
Microbiology	Part 866
Neurology	Part 882
Obstetrical and gynecological	Part 884
Ophthalmic	Part 886
Orthopedic	Part 888
Pathology	Part 864
Physical medicine	Part 890
Radiology	Part 892
Toxicology	Part 862

Table 2.3 Class I and class II exempt devices [29,30].

Part 610	General biological product standards
Part 660	Additional standards for diagnostics substances for laboratory tests
Part 862	Clinical chemistry and clinical toxicology devices
Part 864	Hematology and pathology devices
Part 866	Immunology and microbiology devices
Part 868	Anesthesiology devices
Part 870	Cardiovascular devices
Part 872	Dental devices
Part 874	Ear, nose, and throat devices
Part 876	Gastroenterology—urology devices
Part 878	General and plastic surgery devices
Part 880	General hospital and personal use devices
Part 882	Neurological devices
Part 884	Obstetrical and gynecological devices
Part 886	Ophthalmic devices
Part 888	Orthopedic devices
Part 890	Physical medicine devices
Part 892	Radiology devices

to other important health related areas. Under this act, class II devices are exempted from premarket notification submissions; however, the device manufacturers have to comply with FDA's Good Manufacturing Practices (GMP) requirements.

2.6 Emergency use authorization

Emergency use authorization (EUA) is a special approval that the FDA grants to medical device and pharmaceutical drug makers during public health emergencies to protect against chemical, biological, radiological, and nuclear (CBRN) threats. Under this authorization, the FDA may allow unapproved medical products or unapproved uses of approved medical products to be used in emergency when there are no approved alternatives [31]. For example, the FDA approved use of Pfizer/BioNTech, Moderna, and Johnson & Johnson vaccines to treat SARS-CoV-2 (COVID-19) in 2020 and 2021 [32].

The FDA was granted authority to regulate medical devices by the Congress in 1976. The Center for Devices and Radiological Health (CDRH) oversees the review and approval process if the device does not include biologic component. However, if the device includes a biologic component then the Center for Biologics Evaluation and Research (CBER) may oversee its evaluation [31].

Definition of the device is defined in section 201(h) of the Food, Drug & Cosmetic Act (FD&C Act) as:

an instrument, apparatus, implement, machine, contrivance, implant, in-vitro reagent, or other similar or related article, including a component part or accessory

Device must meet following standards [17]:
- It should be recognized in the official National Formulary, or in the United States Pharmacopoeia.
- Used in the diagnosis of disease or other conditions, or in the cure, mitigation, treatment, or prevention of disease in humans or other animals.
- Affect the structure or any function of the body of humans or animals.
- Does not achieve its primary intended purposes through chemical action within or on body of humans or animals.

2.7 Premarket notification, 510(k)

Premarket notification, 510(k), submission is required for manufacturers of class I, II, and III devices that do not require PMA and are not exempted from 510(k) requirements. This submission is made to the FDA to demonstrate that the device is safe and effective and equivalent the predicate device [33].

Premarket notification, 510(k), process requires manufacturers to provide substantial equivalence to another legally US marketed device [33]. By substantial equivalence, the FDA requires the manufacturer to show that the new device is safe, effective, and equivalent to the predicate device. A device is substantially equivalent if the intended use is the same as the predicate and the technological characteristics are the same as the predicate. It is also substantially equivalence if the intended use is the same as the predicate but has different technological characteristics and does not raise different questions of safety and effectiveness.

According to the FDA, four types of manufacturers require 510(k) submission [33]:

1. Domestic manufacturers that are introducing a device to the US market
2. Specification developers introducing a device to the US market
3. Repackages or relabelers who make significant change to the labeling
4. Foreign manufacturers or exporters introducing device to the US market

Table 2.4 provides an overview of requirements when 510(k) submission is needed and when it is not [33].

2.8 Premarket approval

The ultimate goal of the FDA is to ensure sufficient controls are in place that can determine safety and effectiveness of the device. Unlike class I and class II devices, class III devices support human life and can cause potential illness and harm to the patient. Due to the level of risk associated with class III devices, the FDA has determined that due to the level of high risk associated with class III devices, general and special controls may not be sufficient to determine safety and effectiveness of the device and additional controls may need to be established. These additional controls are in the form of PMA [34].

PMA is the scientific and regulatory review process with the FDA where the device manufacturer has to demonstrate that the device has met FDA's stringent requirements for safety and effectiveness. The technical section of

Table 2.4 Overview of 510(k) requirements [33].

510(k) needed	510(k) not needed
1. Manufacturers who want to market and sell a device in the United States are required to make 510(k) submission at least 90 days prior to offering the device for sale 2. A change or modification to a legally marketed device and that could significantly affect its safety or effectiveness 3. A change or modifications to an existing device, where the modifications could significantly affect the safety or effectiveness of the device or the device is to be marketed for a new or different intended use 4. If devices or components sold directly to end users as replacement parts 5. If significant changes made to the labeling or otherwise affect any condition of the device	1. Unfinished devices sold to another manufacturer for further processing or selling components to be used in the assembling of devices by other manufacturers 2. Device not being marketed or commercially distributed 3. Device used for developing, evaluating, or testing 4. Devices used for clinical evaluation 5. Distributing another manufacturer's domestically manufactured device 6. If the manufacturer is the repackager or relabeler and the existing labeling condition of the device has not significantly changed 7. Commercially released device that has not significantly changed or modified in design, components, method of manufacture, or intended use 8. Importer of the foreign made medical device and 510(k) has been submitted by the foreign manufacturer and received marketing clearance 9. Device is exempted from 510(k) by regulation (21 CFR 862–892)

the PMA application contains the data and information that will allow the FDA to determine whether to approve or disapprove the application. These sections are usually divided into nonclinical laboratory studies and clinical investigations [34]:

(i) Nonclinical Laboratory Studies Section

The first section is the nonclinical laboratory studies section that includes information on microbiology, toxicology, immunology, biocompatibility, stress, wear, shelf life, and other laboratory or animal tests. Nonclinical studies for safety evaluation must be conducted in compliance with 21 CFR Part 58 (Good Laboratory Practice for Nonclinical Laboratory Studies).

(ii) Clinical Investigations Section

The second section is the clinical investigations section that includes study protocols, safety and effectiveness data, adverse reactions and complications, device failures and replacements, patient information, patient complaints, tabulations of data from all individual subjects, results from statistical analysis, and any other information from the clinical investigations.

2.9 Postapproval market requirements

Upon completion of PMA process, the FDA may impose postapproval requirements in the PMA approval order. Some of the postapproval requirements may include the following [35]:
- General Requirements
- Postapproval Report
- PMA Supplement
- Medical Device Reporting (MDR)
- Postmarket Surveillance Studies

2.9.1 General requirements

The FDA may provide post-PMA general requirements that device manufacturer must comply. Failure to comply with any postapproval requirement constitutes a reason for withdrawing approval of a PMA. Some of these general requirements include the following [35]:
- Restriction of the sale, distribution, or use of the device
- Ongoing evaluation and periodic reporting on the safety, effectiveness, and reliability of the device
- Inclusion of identification codes on the device or its labeling or on cards provided to patients in the case of implants
- Batch testing of the device
- Device tracking requirements
- Submission of reports at specified intervals
- Displaying in the labeling and in the advertising of any restricted device warnings, hazards, or precautions important for safe and effective use
- Maintenance of records that will enable the applicant to submit to FDA information needed to trace patients if such information is necessary to protect the public health
- Maintenance of records for specified periods of time and organization and indexing of records into identifiable files to enable FDA to determine whether there is reasonable assurance of the continued safety and effectiveness of the device

- Grant access to FDA to access records and reports permit authorized FDA employees to copy and verify such records and reports.
- Permit authorized FDA employees to inspect manufacturing facilities to verify that the device is being manufactured, stored, labeled, and shipped under approved conditions
- Device not to be manufactured, sterilized, packaged, stored, labeled, distributed, or advertised if not specified in the PMA order for the device

2.9.2 Postapproval report

Submission of postapproval reports is required by the FDA for continued approval of the PMA. Postapproval reports are annual reports that are submitted to FDA at intervals of 1 year from the date of approval of the original PMA. The report should include the beginning and ending date of the period covered by the report and following information [35]:

1. Identify all changes required to be reported to the FDA including changes described in PMA supplement
2. Bibliography and summary of information not previously submitted to FDA as part of the PMA and that is known to the device manufacturer including following:
   a. Unpublished reports of data from any clinical investigations and nonclinical laboratory studies related to the device
   b. Reports in the scientific literature concerning the device

The FDA will then review the summary and bibliography and make a determination whether the agency needs a copy of unpublished or published reports.

The annual report should contain information related to the original PMA and any subsequent PMA supplements. In addition, postapproval reports for PMA supplements approved under the original PMA are to be included in the next and subsequent annual reports for the original PMA unless specified otherwise in the approval order for the supplement.

2.9.3 Premarket supplement

PMA supplement is an additional documentation that FDA requires the device manufacturer to submit before making any changes that affect safety and effectiveness of the device. However, an exception may be granted for a type of change for which "Special PMA-Changes Being Effected" [35].

A PMA supplement must be submitted:
- When unanticipated adverse effects, increases in the incidence of anticipated adverse effects, or device failures necessitate a labeling, manufacturing, or device modification.
- When the device is to be modified and the modified device should be subjected to animal or laboratory or clinical testing designed to determine if the modified device remains safe and effective.

2.9.4 Medical Device Reporting

The holder of an approved PMA must comply with all applicable postmarket requirements required by device regulations. Medical device reporting requirements should comply with 21CFR803, Medical Device Reporting [35]. The PMA holder must:
- Report deaths and serious injuries that a device has or may have caused or contributed to
- Establish and maintain adverse event files

2.9.5 Postmarket surveillance studies

The FDA may require the device manufacturer to conduct a postmarket surveillance as a condition of approval if one of the following situation(s) is met after their assessment [35]:
1. Failure of the device would likely cause serious adverse health consequences.
2. The device is intended to be implanted in the human body for more than 1 year.
3. The device is expected to be a life sustaining or life supporting device used outside a device user facility.
4. The device is expected to have significant use in pediatric populations.

On receiving an order to conduct postmarket surveillance study from FDA, manufacturers must submit a plan for approval within 30 days. The FDA will have 60 days from the time they receive manufacturer's proposed plan to determine if the person designated to conduct the surveillance is qualified and experienced, and if the plan will collect useful data necessary to protect the public health [35].

2.10 FDA's Accelerated Approval Program

The objective of FDA's Accelerated Approval Program is to provide early approval of drugs that can treat serious illness based on a surrogate endpoint. Results from surrogate endpoint assessments are likely to be available prior to the trial endpoints that may show true clinical benefits.

A surrogate endpoint is a marker, such as a laboratory measurement, radiographic image, physical sign, or other measure that is thought to predict clinical benefit, but is not itself a measure of clinical benefit [36−38]. By using surrogate endpoints to demonstrate efficacy, the accelerated approval program can help provide earlier drug access to patients with severe disease states where there is an unmet need [37,39].

Once the manufacturer receives accelerated approval, they are required to perform phase four clinical trials, to confirm clinical benefit. If the phase four trial results are positive, the drug goes through the regular approval; however, if the results are negative, the FDA has the option to reevaluate or completely remove the drug from the market [37−39].

2.11 FDA's Breakthrough Devices Program

Breakthrough Devices Program is a voluntary program created by the FDA to highlight certain medical devices that provide more effective treatment or diagnosis of life-threatening or irreversibly debilitating diseases or conditions [40]. This program replaces the Expedited Access Pathway and Priority Review for medical devices.

The objective of the Breakthrough Devices Program is to provide timely access of medical devices to patients and medical professional by speeding up their development, assessment, and review while retaining necessary standards that are required for premarket approval, 510(k) clearances, and de novo marketing authorization.

2.11.1 Benefits

The benefit of the Breakthrough Devices Program is that it provides an opportunity to manufacturers to interact with the FDA experts to address range of topics during the premarket review phase. This interactive approach help manufacturers receive feedback from FDA, identify areas of agreement in a timely way, and can also expect prioritized review of their submission [40,41].

2.11.2 Program principles

The Breakthrough Devices Program comprises two phases:
- Designation Request Phase
- Device Development and Review Phase

In the Designation Request Phase, an interested sponsor of the device requests the FDA to grant the device Breakthrough Device designation. In the second phase, actions are taken to expedite development of the device

and the prioritized review of regulatory submissions. The FDA developed guiding principles that describe the philosophy of the Breakthrough Devices Program [41]. They are as follows:
- Interactive and Timely Communication
- Pre- and Postmarket Balance of Data Collection
- Efficient and Flexible Clinical Studies
- Review Team Support
- Senior Management Engagement
- Priority Review
- Manufacturing Considerations for PMA Submissions

2.11.3 Program features

The novelty of a Breakthrough Device can present key challenges because both the sponsor and FDA may face more uncertainty about how best to evaluate the device's safety and effectiveness. To expedite the development of these devices, FDA intends to offer sponsors a menu of options as shown below that offer opportunities for early and regular interaction with the FDA [41]:
- Breakthrough Device Sprint Discussion
- Data Development Plan
- Clinical Protocol Agreement
- Other Presubmission for Designated Breakthrough Device
- Regular Status Updates

References

[1] Food and Drug Administration (FDA). https://www.fda.gov/regulatory-information/laws-enforced-fda.
[2] Medical Devices: Full Definitions, World Health Organization, Geneva. https://www.who.int/medical_devices/full_deffinition/en/.
[3] L.R. Atles, A Practicum for Biomedical Engineering and Technology Management Issues, Kendall Hunt Publishing, Dubuque, 2008.
[4] Medical Device Timeline, Morgridge Institute for Research, Madison WI. https://morgridge.org/outreach/teaching-resources/medical-devices/medical-devices-timeline/.
[5] R. Fenton, What are the Differences in the FDA Medical Device Classes, Qualio Blog, Quality & Compliance Hub, June 17, 2021. https://www.qualio.com/blog/fda-medical-device-classes-differences.
[6] Classifying Medical Devices, Food and Drug Administration (FDA). https://www.fda.gov/medical-devices/overview-device-regulation/classify-your-medical-device.
[7] Overview of Medical Device Classification and Reclassification, Food and Drug Administration (FDA). https://www.fda.gov/about-fda/cdrh-transparency/overview-medical-device-classification-and-reclassification.
[8] Wikipedia. https://en.wikipedia.org/wiki/Toothbrush.

[9] E. Schulz, Syringe, Wikimedia Commons, Free Media Repository. https://commons.wikimedia.org/wiki/File:Syringe.jpg.
[10] Essentio MRI Pacemakers, Boston Scientific. https://www.bostonscientific.com/en-US/products/pacemakers/accolade-mri-and-essentio-mri-pacemakers.html.
[11] Wikipedia. https://simple.wikipedia.org/wiki/Defibrillation.
[12] Wikipedia. https://en.wikipedia.org/wiki/Electric_toothbrush.
[13] Wikipedia. https://en.wikipedia.org/wiki/Tongue_depressor.
[14] Wikipedia. https://en.wikipedia.org/wiki/Scalpel.
[15] 3M Nexcare No Hurt Wrap. https://www.nexcare.com/3M/en_US/nexcare/products/catalog/~/Nexcare-No-Hurt-Wrap/?N=4326+3294529207+3294631804&rt=rud.
[16] Stryker. https://www.stryker.com/us/en/acute-care/products/spirit-select.html.
[17] American Academy of Orthopedic Surgeons. https://www.aaos.org/aaosnow/2020/jul/research/research03/.
[18] Wikipedia. https://en.wikipedia.org/wiki/File:Telehealth_-_Blood_Pressure_Monitor.jpg.
[19] Wikipedia. https://en.wikipedia.org/wiki/Pregnancy_test.
[20] Wikipedia. https://en.wikipedia.org/wiki/Syringe.
[21] Wikipedia. https://en.wikipedia.org/wiki/Contact_lens.
[22] Medical Gloves for Covid-19, Food and Drug Administration (FDA). https://www.fda.gov/medical-devices/coronavirus-covid-19-and-medical-devices/medical-gloves-covid-19.
[23] What to Know About Breast Implants, Food and Drug Administration (FDA). https://www.fda.gov/consumers/consumer-updates/what-know-about-breast-implants.
[24] FDA Lifts Injunction on Philips defibrillators, Diagnostic and Interventional Cardiology, April 24, 2020. https://www.dicardiology.com/content/fda-lifts-injunction-philips-defibrillators.
[25] MED-EL Cochlear Implant System – P000025/S104, Food and Drug Administration (FDA). https://www.fda.gov/medical-devices/recently-approved-devices/med-el-cochlear-implant-system-p000025s104.
[26] Wikipedia. https://en.wikipedia.org/wiki/Hip_replacement.
[27] FDA Medical Device Classification: Process Explained, Quasar Medical, May 14, 2019. https://quasar-med.com/fda-medical-device-classification/.
[28] Device Classification Panels, Food and Drug Administration (FDA). https://www.fda.gov/medical-devices/classify-your-medical-device/device-classification-panels.
[29] Class I, II and III Exemptions, Food and Drug Administration (FDA). https://www.fda.gov/medical-devices/classify-your-medical-device/class-i-ii-exemptions.
[30] Food and Drug Administration. https://www.accessdata.fda.gov/scripts/cdrh/cfdocs/cfpcd/315.cfm.
[31] Emergency Use Authorization, Food and Drug Administration (FDA). https://www.fda.gov/emergency-preparedness-and-response/mcm-legal-regulatory-and-policy-framework/emergency-use-authorization.
[32] Johnson &Johnson Emergency Use Authorization, Food and Drug Administration (FDA). https://www.fda.gov/news-events/press-announcements/fda-issues-emergency-use-authorization-third-covid-19-vaccine.
[33] Premarket Notification 510(k), Food and Drug Administration (FDA). https://www.fda.gov/medical-devices/premarket-submissions/premarket-notification-510k.
[34] Premarket Approval (PMA), Food and Drug Administration (FDA). https://www.fda.gov/medical-devices/premarket-submissions/premarket-approval-pma.
[35] PMA Post-Approval Requirements, Food and Drug Administration (FDA). https://www.fda.gov/medical-devices/premarket-approval-pma/pma-postapproval-requirements.

[36] Accelerated Approval Program, Food and Drug Administration (FDA). https://www.fda.gov/drugs/information-health-care-professionals-drugs/accelerated-approval-program.
[37] R. Chandanais, A look at the FDA's accelerated approval program, Pharm. Times (November 16, 2017). https://www.pharmacytimes.com/contributor/ryan-chandanais-ms-cpht/2017/11/a-look-at-the-fdas-accelerated-approval-program.
[38] U.S. Department of Health and Human Services. Guidance for Industry. Expedited Programs for Serious Conditions — Drugs and Biologics. google.ca/url?sa=t&rct=j&q=&esrc=s&source=web&cd=4&cad=rja&uact=8&ved=0ahUKEwid_LrVnLzXAhVE2SYKHTjBCmwQFgg3MAM&url=https%3A%2F%2Fwww.fda.gov%2Fdownloads%2FDrugs%2FGuidances%2FUCM358301.pdf&usg=AOvVaw2Hk_NBPGLiGXDYVSiu8eLb. (Accessed 13 November 2017). Published May 2014.
[39] H. Naci, K.R. Smalley, A.S. Kesselheim, Characteristics of pre-approval and post-approval studies for drugs granted accelerated approval by the US Food and Drug Administration, J. Am. Med. Assoc. 318 (7) (2017) 626–636, https://doi.org/10.1001/jama.2017.
[40] Breakthrough Devices Program, Food and Drug Administration (FDA). https://www.fda.gov/medical-devices/how-study-and-market-your-device/breakthrough-devices-program.
[41] Guidance Documents, Breakthrough Devices Program, Food and Drug Administration (FDA). https://www.fda.gov/regulatory-information/search-fda-guidance-documents/breakthrough-devices-program.

CHAPTER THREE

Selection of materials for construction of medical devices

Contents

3.1 Overview of materials	45
3.1.1 Plastics	45
3.1.2 Elastomers	46
3.1.3 Thermosets	47
3.1.4 Metals	47
3.1.5 Glass	49
3.1.6 Ceramics	49
3.2 Material requirements for medical device qualification	49
3.2.1 Biocompatibility	49
3.2.1.1 Toxicological risk assessment	53
3.2.2 Extractables and leachables	56
3.2.3 Sterilization	57
3.2.3.1 Steam (autoclave) sterilization	58
3.2.3.2 Gamma radiation	60
3.2.3.3 Ethylene oxide sterilization	60
3.2.3.4 Dry heat sterilization	61
References	63

3.1 Overview of materials

3.1.1 Plastics

Plastics play a significant role in revolutionizing healthcare. Because of their durability and easy-to-sanitize characteristics, they are used in medical devices. Plastics have proved to be one of the few versatile materials that have been able to adapt along with the dynamic nature of the industry. For example, disposable plastic syringes, blood bags, new heart valves, and other medical devices are some of the many ways plastics have been used. Some of their advantages over other materials include the following [1,2]:
- Infection resistance
- Improved comfort and safety

Applications of Polymers and Plastics in Medical Devices
ISBN: 978-0-12-820980-6
https://doi.org/10.1016/B978-0-12-820980-6.00003-5

- Low cost
- Innovation

Infection resistance

Plastics can be used as an antimicrobial material to create surfaces that can repel or kill bacteria. These antimicrobial plastics are known to have 99.99% effectiveness in killing bacteria and other microbes even in an environment where surfaces are not cleaned regularly. Examples include devices such as surgical gloves, syringes, insulin pens, IV tubes, catheters, and inflatable splits [1].

Improved comfort and safety

Plastics can be used as a hypoallergenic material on patients that are susceptible to allergies. Because they are durable, plastics can be used in medical safety devices such as tamper-proof caps on medical packaging, blister packs, and medical waste disposal bags. Plastics are shatter-proof and are used in storage and transportation industry. They are impermeable material and are used in biohazard bags to transport medical waste.

Traditionally, metal has been the preferred material for medical devices used as prosthetics. However, plastics because of their durability and versatility are now used as an alternate material of choice to metal components and are able to alleviate pain and improve patient comfort [1].

Low cost

Plastics are a unique material that can not only be mass-produced at cost-effective rates but also allow for a wider range of applications. Unlike metal-based medical devices, certain plastics such as polyamides are not susceptible to wearing and corrosion and are able to withstand applied stress. This reduces additional cost that is involved in maintaining quality products.

Innovation

Plastics are very versatile material and can be manufactured according to specific application. They can be extruded into films or injection molded into complex medical devices. They are being used as cardiovascular and surgical devices such as vascular grafts, pacemakers, stents, and catheters.

3.1.2 Elastomers

Elastomers are form of polymers with rubbery properties. Materials such as silicones, thermoplastic elastomers, thermoplastic polyurethanes (TPUs), silicone elastomers, and hydrogels have found their use in medical device industry because of their durability, biocompatibility, design flexibility, and favorable performance to cost ratios. Some of the applications include cardiovascular devices, prosthetic devices, general medical care products, transdermal therapeutic systems, orthodontics, and ophthalmology [3—6].

3.1.3 Thermosets

Thermosets are materials which remain in a permanent solid state after being cured. During the curing process, polymer molecules get cross-linked and form irreversible bond. As a result, these materials cannot be melted, processed, or reprocessed even at extremely high temperatures. Epoxy, silicone, polyurethane, and phenolics are some of the examples. Some of the advantages include the following:
- Because of its low viscosity, thermosets are easy to work since they exist in liquid form at room temperature
- No toxic fumes generated during processing
- Molding processes use less heat and pressure
- Excellent flowability
- Dimensionally stable
- Little or no shrinkage
- Excellent resistance toward environmental factors
- Resistant to corrosion and chemicals

Thermosets are limited in where they can be used based on their properties. In medical device industry, thermosets are used in orthopedic and dental applications. Examples include housing for dental, surgical, and orthopedic handheld medical devices.

3.1.4 Metals

Metals and metal alloys are commonly used in implanted medical devices and in inserts. These materials are sometimes in contact with parts of the body for extended periods of time. The most common metals and alloys used in implants include niobium, tantalum, nitinol, copper, stainless steel, and titanium. Other metals, such as cobalt—chrome alloy, gold, platinum, silver, iridium, and tungsten, are also common in many medical devices. Table 3.1 shows types of metal used in medical devices, their properties, and applications [7,8].

Table 3.1 Metals used in medical devices [7,8].

Material	Properties	Application
Niobium	• Corrosion resistant • Physiologically inert • Extremely high melting point	• Pacemakers
Tantalum	• High ductility • Corrosion resistant • Easily welded • Good dielectric properties	• Bone implants • Vascular clips • Flexible stents • Diagnostic marker bands • Catheter plastic compounding additive
Nitinol	• Shape memory alloy with superelastic properties • Physiologically compatible with human body • Chemically compatible with human body • Light weight	• Stents • Heart valve tools • Staples • Bone anchors • Septal defect devices • Diagnostic guidewires • Arch wires for braces • Biopsy devices • Repositionable wire markers
Copper	• Outstanding antiviral and antibacterial • High conductivity	• Antigerm surfaces • Medical electric devices • Dental implants • Biocidal • Shielded metal wires and strips
Stainless steel	• High corrosion resistance • Low carbon content • Does not react with bodily tissue • Sterilizable • Wear resistant	• Medical appliances • Orthopedic replacements such as replacement hip joints, or to stabilize broken bones with screws and plates • Tweezers • Forceps • Hemostats
Titanium	• Light • Excellent biocompatibility • Outstanding corrosion resistance	• Dental implants • Surgical devices • Pacemaker cases • Hip replacements

3.1.5 Glass

Glass fibers bring many advantages over other materials when it comes to light delivery. Endoscopes and dental instruments are two such examples. The following properties make glass fibers stand out among other materials [9].
- (a) Stable at temperatures up to 350°C
- (b) Can withstand autoclave sterilization, which makes glass ideal for light delivery in reusable devices such as dental instruments or endoscopes
- (c) Superior light performance
- (d) Strong and bendable

3.1.6 Ceramics

Ceramic has been used in different applications for a long time; however, there has been recent drive for it to be used in medical processes and applications. Ceramic has high melting point, low electrical and thermal conductivity, and inertness to chemicals. It is typically used in surgical implants, prosthetics, and various medical tools and devices [10]. Some common applications include the following:
- X-ray tubes
- Pressure sensors
- Dental screws and bridges
- Femoral head implants for hip replacement
- Hand tools
- Valves
- Filler
- Femoral balls in hip replacements
- Dental brackets
- Neurostimulators

3.2 Material requirements for medical device qualification

3.2.1 Biocompatibility

Biocompatibility standard, ISO 10993-1 from the International Organization for Standardization (ISO) has been adopted for applications in the United States and around the globe. The FDA has also developed its own guidance document to further clarify the use of ISO 10993 and incorporates risk-based approach to determine if biocompatibility testing is needed [11].

The ISO 10993 Part 1 uses an approach to biological test selection for device materials that includes seven principles [11]. The selection of device materials to be tested and the toxicological evaluations should consider full characterization of all materials of manufacture. For proprietary materials, device master files (MAF) can be submitted to FDA to assist in determining the formulation of some components of the final device.

1. The materials of manufacture, the final product, and possible leachable chemicals or degradation products should be considered for their relevance to the overall toxicological evaluation of the device.
2. Tests to be utilized in the toxicological evaluation should consider the bioavailability of the material (i.e., nature, degree, frequency, duration, and conditions of exposure of the device to the body).
3. Any in vitro or in vivo experiments or tests should be conducted in accordance with recognized good laboratory practice (GLP).
4. Full experimental data should be submitted to the reviewing authority unless testing is conducted according to a recognized standard that does not require data submission.
5. Any change in chemical composition, manufacturing process, physical configuration, or intended use of the device should be evaluated with respect to possible changes in toxicological effects and the need for additional toxicity testing.
6. The toxicological evaluation performed in accordance with this guidance should be considered in conjunction with other information from other nonclinical tests, clinical studies, and postmarket experiences for an overall safety assessment.

Manufacturers bringing new medical devices to market have to meet ISO 10993-1 requirements. The FDA has issued additional recommendations for medical device manufacturers that require evaluating biocompatibility of the medical device materials and processes using the biological evaluation process (BEP). BEP process includes developing biological evaluation plan, performing testing and risk assessment, and finalizing the biological evaluation report [12,13].

A biological evaluation plan (BEP) is an initial risk assessment to review device materials, identify potential risks, and suggest possible evaluations and testing to address the risks identified based on the nature of patient contact of the device. This serves as an initial risk assessment outlined in ISO 10993-1 and provides good internal documentation of the approach used to address biocompatibility. BEP takes into consideration the materials, processing, and historical use of the device. At the end of the evaluation, results of all tests and written evaluations are summarized in a Biological Evaluation Report (BER) that is submitted along with test results to the FDA.

Fig. 3.1 shows a flowchart providing guidance for materials that require biocompatibility evaluation [13]. Additional testing may be needed if new materials or processes are used.

A key factor in this process is to define the level of device contact with the body, i.e., direct versus indirect. All biocompatibility requirements are met for devices that make no contact with the body. For devices that contact

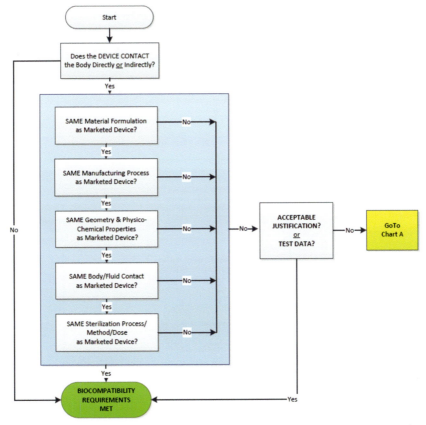

Figure 3.1 Flowchart showing biocompatibility evaluation plan [13]. *Source: Food & Drug Administration, Republished on 20th December 2021.*

Applications of Polymers and Plastics in Medical Devices

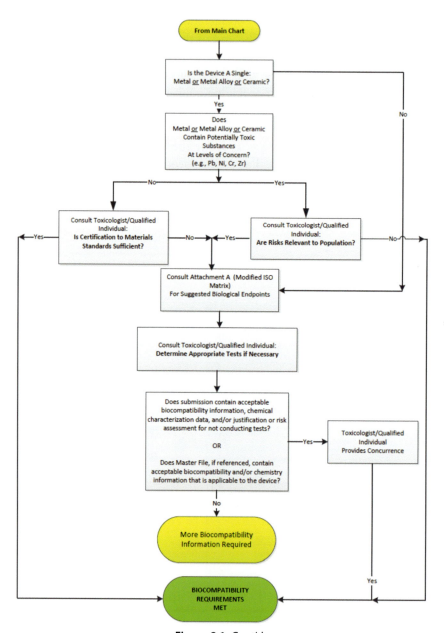

Figure 3.1 Cont'd.

the body, the material has to undergo biocompatibility evaluation according to the end point guidance shown in Appendix A of the FDA guidance (Table 3.2) [13].

The selected test program and biological endpoints depend on several factors, including time duration of contact with the device. If a component is subjected to body tissue or fluid contact, it is important to know the duration of the contact. The contact is divided into three types—(1) limited when the contact is less than 24 h, (2) prolonged when the contact is greater than 24 h and less than 30 days, and (3) permanent when the contact is greater than 30 days [12,13].

The specific biological tests required are based on chemical characteristics of device materials, and the nature, degree, frequency, and duration of exposure to the body. These tests may include the following:
- In vitro cytotoxicity
- Acute, subchronic, and chronic toxicity
- Irritation
- Sensitization
- Hemocompatibility
- Implantation
- Genotoxicity
- Carcinogenicity

3.2.1.1 Toxicological risk assessment

The toxicological risk assessment (TRA) is an important tool in the safety assessment of medical devices, providing a chemical-based approach. It is the process of gathering all possible toxicity data about the materials of construction, processing materials and potential contaminants, and using this information to provide a risk profile [14,15].

TRA mainly comprised four activities [16]:
- Hazard identification and data evaluation
- Dose—response assessment
- Exposure assessment
- Risk characterization

Hazard identification and data evaluation
The objective of this step is to identify the types of adverse health effects that can be caused by exposure to questionable agents and to characterize the

Table 3.2 Biocompatibility evaluation endpoints considerations [13].

Medical device categorization by			Biological effect											
Nature of body contact		Contact duration A—limited (≤24 h) B—prolonged (>24 h to 30 d) C—permanent (>30 d)	Cytotoxicity	Sensitization	Irritation or intracutaneous reactivity	Acute systemic toxicity	Material mediated pyrogenicity	Subacute/ Subchronic toxicity	Genotoxicity	Implantation	Hemo-compatibility	Chronic toxicity	Carcino-genicity	Reproductive/ Developmental toxicity	Degradation
Category	Contact														
Surface device	Intact skin	A	X	X	X										
		B	X	X	X										
		C	X	X	X										
	Mucosal membrane	A	X	X	X										
		B	X	X	X	O	O	O							
		C	X	X	X	O	O	X	X	O		O			
	Breached or compromised surface	A	X	X	X	O	O	O							
		B	X	X	X	O	O	X		O					
		C	X	X	X	O	O	X	X	O		O	O		
External communicating	Blood path, indirect	A	X	X	X	X	X	O			X				
		B	X	X	X	X	X	X			X				
		C	X	X	O	X	X	X	X	O	X	O	O		
	Tissue/bone/ dentin	A	X	X	X	O	O		X						
		B	X	X	X	X	X	X	X	X					
		C	X	X	X	X	X	X	X	X		O	O		
	Circulating blood	A	X	X	X	X	X	X	O	X	X				
		B	X	X	X	X	X	X	X	X	X				
		C	X	X	X	X	O	X	X	X	X	O	O		
Implant device	Tissue/bone	A	X	X	X	X	X	X	X	X					
		B	X	X	X	X	X	X	X	X					
		C	X	X	X	X	O	X	X	X		O	O		
	Blood	A	X	X	X	X	X	X	O	X	X	O			
		B	X	X	X	X	O	X	X	X	X	O	O		
		C	X	X	X	X	O	X	X	X	X	O	O		

O, Additional FDA-recommended endpoints for consideration; X, ISO 10,993 recommended endpoints for consideration.
Source: Food & Drug Administration, Republished on 20th December 2021.

quality and weight of evidence supporting this identification. This process determines whether exposure to a stressor can cause an increase in the incidence of specific adverse health effects (e.g., cancer, birth defects). It is also whether the adverse health effect is likely to occur in humans [16].

Key components of hazard identification involve evaluating the weight of evidence regarding a chemical's potential to cause adverse human health effects.

Dose response
A dose—response relationship describes how the likelihood and severity of adverse health effects (the responses) are related to the amount and condition of exposure to an agent (the dose provided). As the dose increases, the measured response also increases. At low doses, there may be no response. At some level of dose, the responses begin to occur in a small fraction of the study population or at a low probability rate. Both the dose at which response begins to appear and the rate at which it increases given increasing dose can be variable between different pollutants, individuals, exposure routes, etc. [16].

Exposure assessment
It is the process of measuring or estimating the magnitude, frequency, and duration of human exposure to an agent in the environment or estimating future exposures for an agent that has not yet been released. An exposure assessment includes discussion of the size, nature, and types of human populations exposed to the agent, as well as discussion of the uncertainties in the aforementioned information. Exposure can be measured directly but more commonly is estimated indirectly through consideration of measured concentrations in the environment, consideration of models of chemical transport, and fate in the environment and estimates of human intake over time [16].

Risk characterization
A risk characterization conveys the risk assessor's judgment as to the nature and presence or absence of risks, along with information about how the risk was assessed, where assumptions and uncertainties still exist, and where policy choices will need to be made. Risk characterization takes place in both human health risk assessments and ecological risk assessments. A good risk characterization will restate the scope of the assessment, express results

clearly, articulate major assumptions and uncertainties, identify reasonable alternative interpretations, and separate scientific conclusions from policy judgments [16].

3.2.2 Extractables and leachables

Extractables and leachables are trace amount of chemicals that can be extracted from medical devices made from plastics that come in contact with solvents such as alcohol at various temperature of use and storage. Extractables slightly differ from leachables in definition. The difference between extractable and leachable is slight; leachable is a worst-case scenario as it results from more aggressive exposure [17].

The following steps are to be followed to have a successful extractables and leachables program (E/L):

Step 1—Initiate an extractables and leachables program (E/L) which requires understanding the manufacturing process from start to finish as well as knowing and identifying product contact with any liquid.

Step 2—Once all materials that have product contact are known, an extractables and leachables risk assessment can be performed. The following parameters are to be assessed [18]:
1. Material compatibility
2. Proximity of a component to the final product
3. Product composition
4. Surface area
5. Time and temperature
6. Pretreatment steps

Step 3—If additional investigation of extractables and leachables is required, an extractables profile should be obtained from either the vendor or another source. Ideally, this profile should be produced using appropriate methods so that most potential leachables are identified.

Step 4—Perform extraction with at least two solvents.

Step 5—Methods of analysis used for extractables analysis should detect and identify specific, individual, extractable compounds. HPLC with an ultraviolet (HPLC-UV) or mass spectrometer (LC-MS) detector and GC-MS are the most scientifically robust methods for this purpose.

Step 6—Methods such as total organic carbon (TOC), ash, and nonvolatile residue (NVR) analysis can be used individually or collectively to estimate amounts of extractable material present.

Step 7—The identity of certain extractables should be verified by comparing results with commercial analytical standards. If high-quality extractable data are available, a toxicity assessment is performed for leachables based on the extractable profile. This would be considered a worst-case scenario because the concentration and number of extractables are expected to be higher than what will be measured during leachables testing.

3.2.3 Sterilization

Medical devices produced under standard manufacturing conditions in accordance with the requirements for quality management systems can, prior to sterilization, have microorganisms on them. Such products are nonsterile. The purpose of sterilization is to inactivate the microbiological contaminants and thereby transform the nonsterile products into sterile ones [19].

The concept of what constitutes "sterile" is measured as a probability of sterility for each item to be sterilized. This probability is commonly referred to as the sterility assurance level (SAL) of the product and is defined as the probability of a single viable microorganism occurring on a product after sterilization.

SAL is normally expressed a 10^{-n}. For example, if the probability of a spore surviving were one in one million, the SAL would be 10^{-6}. In short, an SAL is an estimate of lethality of the entire sterilization process and is a conservative calculation. Dual SALs (e.g., 10^{-3} SAL for blood culture tubes, drainage bags; 10^{-6} SAL for scalpels, implants) have been used in the United States for many years, and the choice of a 10^{-6} SAL was strictly arbitrary and not associated with any adverse outcomes (e.g., patient infections) [20].

Medical devices are sterilized in a variety of ways including following [21]:
- Steam
- Dry heat
- Gamma radiation
- Ethylene oxide

Medical devices that have contact with sterile body tissues or fluids are considered critical items. These items should be sterile when used because any microbial contamination could result in disease transmission. Such items include surgical instruments, biopsy forceps, and implanted medical devices.

If these items are heat resistant, the recommended sterilization process is steam sterilization, because it has the largest margin of safety due to its reliability, consistency, and lethality. However, reprocessing heat- and moisture-sensitive items requires use of a low-temperature sterilization technology (e.g., ethylene oxide, hydrogen peroxide gas plasma, peracetic acid) [20].

3.2.3.1 *Steam (autoclave) sterilization*

Steam sterilization is a process where a product is exposed to saturated steam under pressure killing microorganisms by reducing the time and temperature required to denature or coagulate proteins in the microorganisms [22].

Steam sterilization cycles generally have three phases [23]:

(a) Preconditioning

During this phase, air is removed from the chamber and the load is humidified by means of alternating vacuum and pressure pulses.

(b) Exposure

During this phase, the chamber temperature is raised to and held at the programmed sterilizing temperature for the programmed exposure time.

(c) Postconditioning

During this phase, dry goods loads are cooled and dried or a liquids load is cooled. The chamber pressure is brought to atmospheric.

Benefits of steam sterilization

Steam sterilization has many benefits as a sterilization method in healthcare facilities [24]:

- Low cost, safety, and efficacy
- Speed and productive
- Range of products can be processed in a steam sterilizer including surgical instruments, implantable medical devices, and surgical linens

Steam sterilization process

There are seven main parameters in steam sterilization process:

- Air removal
- Drying
- Steam contact
- Time
- Temperature
- Pressure
- Moisture

Air removal Air is removed from the sterilizer chamber to an absolute minimum level and the product is loaded prior to operation to secure saturated steam conditions.

Drying The load is dry when leaving the sterilizer to prevent recontaminated. Proper drying is normally performed by applying a vacuum to the chamber at the end of the cycle, which boils all condensates and transports them away through the vacuum system.

Steam contact The direct steam contact to the potentially contaminated surface is maximized to ensuring stored energy is transferred to the product by the means of condensing.

Time Time is a critical factor that determines minimum duration needed to kill surface spores. It is closely linked with temperature as both parameters determine the killing effect. The killing effect is defined as the lethal value, which should reach the same value by sterilizing at 121°C for 15 min as one would by sterilizing at 134°C for 3 min. The SAL must be considered when choosing the required lethality value for a specific application. The SAL required differs based on the application, but it is typically defined as sterile around 1/1,000,000, which means that only one out of a million bacteria will have survived the sterilization process.

Temperature As temperature is linked to the lethality value directly, this parameter is also used to determine performance of the autoclave.

Pressure To avoid risk of air pockets, pressure is used to convert values into theoretical temperature, which then can then be compared to the actual temperature and assess whether the steam is saturated.

Moisture Steam moisture has an extremely high impact on destroying proteins by denaturation (coagulation), which is why it is crucial to utilize saturated steam. The steam should be clean, and superheated steam (above its saturation temperature) must be avoided as it will not have enough moisture to ensure proper sterilization if this happens.

Common mistakes in stem sterilization
Common mistakes that are made when performing steam sterilization:
1. Containers with closed valves, empty glass bottles with tightened screw caps, or secured aluminum foil are placed in the sterilizer.
2. Pouched and/or heavily wrapped items are tightly packed in the chamber.
3. Heavier items are placed on top shelves.
4. Load is too dense, or items are positioned incorrectly in the load.
5. Pouches are placed flat on the sterilizer shelves or stacked on top of one another.
6. Liquids in vented containers are placed in a deep pan to catch boil-over (slow exhaust cycle).
7. "Overcooked" media.
8. Using cold water for vacuum pump that is too hot.

9. Load probe is available, but not used.
10. Pressure/vacuum rate control is available, but not used.

3.2.3.2 Gamma radiation
Gamma radiation is a form of electromagnetic radiation—like X-rays, but with higher energy. The primary industrial sources of gamma rays are radionuclide elements such as cobalt 60, which emit gamma rays during radioactive decay. Gamma rays pass readily through plastics and kill bacteria by breaking the covalent bonds of bacterial DNA. They are measured in units called kiloGrays (kGy) [25].

Gamma radiation provides a number of benefits:
- Can be applied under safe, well-defined, and controlled operating parameters
- Does not a heat- or moisture-generating process
- No residual radioactivity after radiation

Gamma radiation sterilization process
Gamma radiation sterilization is performed by exposing the product to a radiation source, typically cobalt 60 isotope, which decomposes into nickel 60 isotope, firing off gamma rays in the process. These gamma rays can penetrate through the entire product, deactivating microorganisms that may be present [26]. The gamma dosage can be measured in using detectors called dosimeters, which enable parametric release.

Below are some key steps:
Step 1: Quantifying mean bioburden and establishing sensitivity to a low radiation dose
Step 2: Higher dosage (>25 kGy) is applied to provide sterility assurance safety margin
Step 3: Determine the target of <10 probability of nonsterile unit (Sterility Assurance Level, SAL) is established

3.2.3.3 Ethylene oxide sterilization
Ethylene oxide (EtO) is a colorless gas that is flammable and explosive. The effectiveness of EtO sterilization depends on four critical parameters. The operational ranges for these parameters are shown in Table 3.3 [27]. Within certain limitations, an increase in gas concentration and temperature may shorten the time necessary for achieving sterilization.

Advantages and disadvantages of using EtO sterilization are shown in Table 3.4 [27].

Table 3.3 Operational ranges of EtO parameters [27].

Parameter	Operational range
Gas Concentration	450–1200 mg/L
Temperature	37–63°C
Relative humidity	40%–80%
Exposure time	1–6 h

Table 3.4 Advantages and disadvantages of EtO [27].

Advantages	Disadvantages
• Can sterilize heat- or moisture-sensitive medical equipment without impacting material(s) used in the medical device	• Long cycle time • Cost • Hazardous to patients and staff

Impact of EtO exposure

Acute exposure to EtO may result in irritation to skin, eyes, gastrointestinal or respiratory tracts, and central nervous system depression. Chronic inhalation has been linked to the formation of cataracts, cognitive impairment, neurologic dysfunction, and disabling polyneuropathies. Occupational exposure in healthcare facilities has been linked to hematologic changes and an increased risk of spontaneous abortions and various cancers. EtO should be considered a known human carcinogen [27].

Principles of EtO sterilization

The basic EtO sterilization cycle consists of five stages:
- Preconditioning and humidification
- Gas introduction
- Exposure
- Evacuation
- Air washes

EtO sterilization cycle which does not include aeration time takes 2.5 h to complete. Because EtO is absorbed by many materials, mechanical aeration for 8–12 h at 50–60°C is performed after EtO sterilization cycle is completed to allow desorption of the toxic EtO residual contained in exposed absorbent materials [27,28].

3.2.3.4 Dry heat sterilization

Dry heat sterilization method is only used on materials that get damaged on exposure to moist heat or that are impenetrable to moist heat. Examples include powders, petroleum products, and sharp instrument. Some of the advantages and disadvantages of dry heat sterilization are shown in Table 3.5 [29].

Table 3.5 Advantages and disadvantages of dry heat sterilization method [29].

Advantages	Disadvantages
• Nontoxic and does not harm the environment	• Slow rate of heat penetration and microbial killing
• Relatively low operation cost	• Process is time-consuming
• Penetrates materials	• High temperatures are not suitable for most materials
• Noncorrosive for metals and sharp instrument	

Table 3.6 Time—temperature relationship for sterilization [30].

Time (min)	Temperature °C (°F)
60	170 (340)
120	160 (320)
150	150 (300)

Principles of dry heat sterilization operation

In this process, sterilization is accomplished when the product is exposed to dry heat. Due to the difference in temperature, the heat is then transmitted through rest of the product until the product reaches an equilibrium temperature. Sterilization is achieved when the product maintains exposure at this temperature for a required time. Dry heat oxidizes molecules killing organisms on the surface; however, temperature needs to be maintained for an hour to kill the most difficult of the resistant spores [30].

The time—temperature relationships for sterilization are shown in Table 3.6. It is important that *Bacillus atrophaeus* spores monitored during the dry heat process since these bacteria are more resistant to dry heat than *Geobacillus stearothermophilus* [29,30].

Types of dry heat sterilizers

There are two types of dry heat sterilizers [29]:

(a) Static-air type

The static-air type is referred to as the oven-type sterilizer. In this type of sterilizer, heating coils are in the bottom of the unit that allows the hot air to rise inside the chamber through convection. Since the heat is circulated from the bottom of the unit to the top by convection, these types of sterilizer have poor temperature control, are slower in heating, and require longer time to reach sterilizing temperature when compared with the forced-air type sterilizer.

(b) Forced-air type

In comparison with static-air type, the forced-air sterilizer is a mechanical sterilizer that is equipped with motor-driven blower that circulates heated air throughout the chamber at a high velocity, permitting a more rapid transfer of energy from the air to the equipment being sterilized.

References

[1] M. Naik, 5 Ways Plastics Revolutionized the Healthcare Industry, Medical Product Outsourcing, 2017. https://www.mpo-mag.com/contents/view_online-exclusives/2017-10-09/5-ways-plastics-revolutionized-the-healthcare-industry/.

[2] Advantages of Plastic for the Medical Industry, Avomeen. https://www.avomeen.com/lifesciences-benefits-plastic-medical-industry/.

[3] Shin-Etsu Silicones, Silicone Skin for Prosthetic Liners, Todays Medical Developments. https://www.todaysmedicaldevelopments.com/article/medical-device-shin-etsu-silicone-skin-prosthetics-11015/.

[4] Polymer for Cardiology Applications, Lubrizol Medical. https://www.lubrizol.com/Health/Medical/Markets/Cardiology.

[5] Hydrogels for Medical Devices, Med Device Online. https://www.meddeviceonline.com/doc/hydrogels-for-medical-devices-0001.

[6] H. Haggarty, How Medical-Grade Silicone is Advancing Reliable Patient Care, Plastics Today, 2021. https://www.plasticstoday.com/medical/how-medical-grade-silicone-advancing-reliable-patient-care.

[7] D. Sanchez, Specialty Metals Make Sophisticated Medical Devices Possible, Medical Design Briefs, 2013. https://www.medicaldesignbriefs.com/component/content/article/mdb/features/articles/17206.

[8] C. Williams, The 6 Most Important Metals Used in Medicine, Star Rapid. https://www.starrapid.com/blog/the-6-most-important-metals-used-in-medicine/.

[9] Five Reasons Why Glass is Best for Medical Devices, Schott North America Inc. https://www.us.schott.com/innovation/five-reasons-why-glass-is-best-for-medical-devices/.

[10] Medical Ceramics, Thomas Publishing Company. https://www.thomasnet.com/articles/custom-manufacturing-fabricating/medical-ceramics/.

[11] FDA Biological Testing of Medical Device Materials, Medical Materials Blog, Foster. https://www.fostercomp.com/fda-biological-testing-of-medical-device-materials/.

[12] Food and Drug Administration, Appendix D, "Biocompatibility Evaluation Flow Chart", Use of International Standard ISO 10993-1, Biological Evaluation of Medical Devices — Part 1: Evaluation and Testing within a Risk Management Process — Guidance for Industry and Food and Drug Administration Staff, September 2020.

[13] Food and Drug Administration, Appendix A, Evaluation Endpoints for Consideration, Use of International Standard ISO 10993-1, "Biological Evaluation of Medical Devices — Part 1: Evaluation and Testing within a Risk Management Process" — Guidance for Industry and Food and Drug Administration Staff, September 2020.

[14] Toxicological Risk Assessment (TRA), Toxicon. https://toxikon.com/testing-service/toxicological-risk-assessment-tra/.

[15] Toxicity Risk Assessment, Medical Engineering Technologies. https://met.uk.com/medical-device-testing-services/biocompatibility/toxicity-risk-assessment.

[16] Conducting a Human Health Risk Assessment, United States Environmental Protection Agency. https://www.epa.gov/risk/conducting-human-health-risk-assessment.

[17] L. McKeen, Plastics used in medical devices, in: K. Modjarrad, S. Ebnesajjad (Eds.), Handbook of Polymer Applications in Medicine and Medical Devices, 2014, pp. 32—33.
[18] Extractables and Leachables Subcommittee of the Bio-Process System Alliance, "Recommendations for Extractables and Leachables Testing", BioProcess International, 2008, in: https://bioprocessintl.com/2008/recommendations-for-extractables-and-leachables-testing-183979/.
[19] ISO 11737-2:2019, Sterilization of Health Care Products — Microbiological Methods — Part 2: Tests of Sterility Performed in the Definition, Validation and Maintenance of a Sterilization Process.
[20] Sterilization, Centers for Disease Control and Prevention. https://www.cdc.gov/infectioncontrol/guidelines/disinfection/sterilization/index.html.
[21] Ethylene Oxide Sterilization for Medical Devices, U.S. Food and Drug Administration. https://www.fda.gov/medical-devices/general-hospital-devices-and-supplies/ethylene-oxide-sterilization-medical-devices.
[22] Steam Sterilization for Medical Equipment, Knowledge Center, Steris Healthcare. https://www.steris.com/healthcare/knowledge-center/sterile-processing/steam-sterilization-for-medical-equipment.
[23] M. Dion, W. Parker, Steam sterilization principles, Pharmaceut. Eng. 33 (6) (November/December 2013).
[24] Ellab, The Steam Sterilization Process, AZO Materials, 2019. https://www.azom.com/article.aspx?ArticleID=18733.
[25] J.M. Martin, Understanding gamma sterilization, Biopharm Int. 25 (2) (2012). https://www.biopharminternational.com/view/understanding-gamma-sterilization.
[26] J. Grove, 5 Gamma Radiation Sterilization Tips, Starfish Medical. https://starfishmedical.com/blog/gamma-irradiation-sterilization/.
[27] Ethylene Oxide "Gas" Sterilization, Centers for Disease Control and Prevention. https://www.cdc.gov/infectioncontrol/guidelines/disinfection/sterilization/ethylene-oxide.html.
[28] Ethylene Oxide Sterilization, Sterigenics. https://sterigenics.com/technologies/ethylene-oxide/.
[29] Other Sterilization Methods, Center for Disease Control and Prevention. https://www.cdc.gov/infectioncontrol/guidelines/disinfection/sterilization/other-methods.html.
[30] A. Tankeshwar, Dry Heat Sterilization: Principles, Advantages, Disadvantages, Microbe Online. https://microbeonline.com/dry-heat-sterilization-principle-advantages-disadvantages/.

CHAPTER FOUR

Classification of plastics and elastomers used in medical devices

Contents

4.1 Performance-based selection of plastics and elastomers for medical devices	66
4.1.1 Translation	66
4.1.2 Screening	66
4.1.2.1 Biocompatibility	*66*
4.1.2.2 Drug flow path	*68*
4.1.2.3 Sterilization compatibility	*68*
4.1.2.4 Resistance to hospital cleaners	*68*
4.1.2.5 Mechanical properties	*69*
4.1.2.6 Dimensional stability	*69*
4.1.2.7 Radiopacity	*69*
4.1.2.8 Conductive	*69*
4.1.2.9 Lubrication	*69*
4.1.3 Ranking	70
4.2 Plastics and elastomers used in FDA class I devices	70
4.2.1 Low-performance polymers	70
4.2.1.1 Polypropylene	*71*
4.2.1.2 Polyethylene	*71*
4.2.1.3 Polyvinyl chloride	*71*
4.2.1.4 Polystyrene	*72*
4.2.2 Medium-performance polymers	72
4.2.2.1 Polycarbonate	*72*
4.2.2.2 Polymethyl methacrylate	*73*
4.2.2.3 Polybutylene terephthalate	*73*
4.2.2.4 Polyphenylene oxide	*73*
4.2.2.5 Acrylonitrile butadiene styrene	*74*
4.2.3 High-performance polymers	74
4.2.3.1 Acetal copolymer	*74*
4.2.3.2 Polyether ether ketone	*75*
4.2.3.3 Polyphenyl sulfone	*75*
4.2.3.4 Polysulfone	*75*
4.2.3.5 Polyphenylene sulfide	*75*
4.2.3.6 Polyvinylidene fluoride	*76*
4.2.3.7 Polyetherimide	*76*

Applications of Polymers and Plastics in Medical Devices
ISBN: 978-0-12-820980-6
https://doi.org/10.1016/B978-0-12-820980-6.00009-6

© 2022 Elsevier Inc.
All rights reserved.

	4.2.3.8 Polydimethylsiloxane	76
	4.2.3.9 Thermoplastic polyurethane	77
	4.2.3.10 Thermoplastic elastomer	77
References		77

4.1 Performance-based selection of plastics and elastomers for medical devices

Any material selection process is guided by end-use requirements. As demand for plastic and elastomers in medical devices is increasing, selecting the right material becomes a balancing act between performance requirement, manufacturability, and cost. Although there are other different factors that need careful consideration, the first and the most important is to know how the device will be used. Medical device OEMs can greatly narrow the field of polymer candidates by carefully defining end-use requirements up front and consulting materials supplier early in the design process [1].

Selection of a material requires following given steps:
- Translation
- Screening
- Ranking

4.1.1 Translation

It is important to define key product requirements at this stage. This can be accomplished by asking questions related to exposure to environment, functional, and mechanical considerations [2].

4.1.2 Screening
4.1.2.1 Biocompatibility

It is important to clearly define the application as it becomes easier to find the right material that satisfies performance and regulatory requirements. A key factor in this process is defining the level of body contact the application will be subjected to.

For applications having no contact with bodily fluids and tissues, selection can be made from a wide spectrum of polymers. Material selection will largely be driven by required performance properties, manufacturability, and sterilization requirements. For applications that have contact with the bodily fluids and tissues, the most important material consideration is biocompatibility [1].

Biocompatibility standard, ISO 10993 from the International Organization for Standardization (ISO), has been adopted for applications in the United States and around the globe. The selected test program and biological endpoints depend on several factors, including time duration of contact with the device [3,4]. If a component is subjected to body tissue or fluid contact, it is important to know the duration of the contact. The contact is divided into three types—(1) limited when the contact is less than 24 h, (2) prolonged when the contact is greater than 24 h and less than 30 days, and (3) permanent when the contact is greater than 30 days [2—4].

To identify potential raw material candidates that have a possibility of passing biocompatibility testing, action needs to be taken early during the development phase. The choice of polymer depends on the extent of contact with body fluids, internal and external tissue as prescribed by the regulatory framework [1,2].

Polymer selection based on the application is shown in Table 4.1 [2].

Table 4.1 Polymer selection based on the application [2].

Medical device application	Polymer
Noncontact with human body	- PVC - PA - PE - PS - Epoxy
Short-term contact with human body	- Silicone rubber - Natural rubber - PVC - Polyurethane - PE - PP - Polyester - PEEK - Polyphenylsulfone - Nylon - Teflon - Pebax
Medium term contact with human body	- Nylon - PP - Polyester
Long-term contact with human body	- PE - UHMWPE - PET - Silicone rubber - Polyurethane - PMMA - Polysulfones - Hydrogels polyphosphazenes - Thermoplastic elastomers - Polydimethylsiloxane

4.1.2.2 Drug flow path

Drug flow path also needs to be evaluated early during the development phase. Drug flow path is referred to the path when the material comes in direct contact with the flow of drugs through a device. Chemical resistance, extraction, and biocompatibility between the drug and the raw material need to be evaluated [2].

4.1.2.3 Sterilization compatibility

Sterilization is another important requirement that applies to medical devices. Devices should be evaluated against different sterilization methods. Plastic and elastomeric devices that require body tissue/fluid contact must be evaluated for sterilization, as they react differently to various sterilization methods [2].

There are four common sterilization methods—(a) steam (autoclaving), (b) dry heat, (c) ethylene oxide (EtO), and gamma radiation. Generally, thermoplastics are able to withstand exposure to EtO; however, they get discolored when exposed to high dosage of gamma radiation and eventually, impact their mechanical properties. List of polymers resistant to gamma radiation include ABS, PARA, PEEK, PEI, PES, PSU, PPSU, and TPU. Polycarbonate (PC) has shown that it is resistant to gamma radiation at low dosage; however, it starts to get discolored at higher dosages [2,5].

Steam sterilization or autoclaving is another sterilization method. It uses a combination of heat and moisture for repeated cycles of 3–15 min. Injection-molded parts are prone to high residual stresses that can cause dimensional instability or can warp when exposed to high temperatures of 121–140°C. For this type of sterilization, amorphous polymer tends to be good. Styrenic polymers (ABS, PS) and polyesters (PBT, PET) show poor resistance toward steam sterilization. If the product is not a single-use device, its components will be subjected to multiple sterilizations before being discarded, so plastics which possess superior toughness with a lesser tendency to discolor (PEEK, PEI, PSU, PPSU, and PC) are better choices [2].

4.1.2.4 Resistance to hospital cleaners

Exposure to hospital cleaners such as isopropyl alcohol, bleaches, and peroxides may contribute to deterioration of the plastics and, as a result, can have an impact on performance. Semicrystalline materials such as PP, PE, and polyamides have better chemical resistance than amorphous polymers such as ABS and PC [2].

4.1.2.5 Mechanical properties

When selecting a material, it is important to evaluate mechanical requirements such as strength, stiffness, or impact resistance. Engineered plastics show excellent mechanical properties at low and high temperatures. Other factors that are also evaluated include wear and abrasion resistance and lubricity [2].

4.1.2.6 Dimensional stability

Dimensional stability is another criterion that needs to be evaluated for parts having tight tolerances. This is important since dimensional stability can be impacted by the operating environment such as exposure to chemicals and high temperature. Amorphous polymers have low, uniform shrinkage that provides good dimensional stability and low warpage to hold tight tolerances in complex components as compared with semicrystalline plastics that have higher, less uniform shrinkage but fillers can be used to retard shrinkage [2].

4.1.2.7 Radiopacity

To be able to be seen by X-ray or fluoroscope, certain additives such as barium sulfate can be added. This technique is useful in surgical implants and catheters that need to be seen during surgery [2].

4.1.2.8 Conductive

In some dry powder and aerosol drug delivery applications, static build-up on the surface of thermoplastic part can attract the drug and cause incorrect dosages. Use of permanent antistatic compounds in aerosol drug delivery applications eliminates static build-up and allows accurate dosing [2].

4.1.2.9 Lubrication

Many devices that have any plastic-on-metal or plastic-on-plastic sliding parts need to have high wear resistance. Gears, implants, and sliding covers are all examples of this. UHMWPE shows high wear resistance, and it can be improved through a process called cross-linking, which creates stronger bonding between the molecular chains that make up the polyethylene. Additives such as PTFE and silicone are commonly used to improve lubricity and wear resistance of a plastic material [2].

4.1.3 Ranking

Once screening of the raw material has concluded, it is time to rank all evaluated raw materials. One can plot a Pugh matrix or any other ranking matrix to perform this activity [2].

4.2 Plastics and elastomers used in FDA class I devices

According to the Food and Drug Administration (FDA), class I devices are defined as devices which are [6]:

> not intended for use in supporting or sustaining life or of substantial importance in preventing impairment to human health, and they may not present a potential unreasonable risk of illness or injury.

These devices have minimal contact with patients, are of low risk, and are subjected to the fewest regulatory requirements. Some examples of class I devices include tongue depressor, oxygen mask, reusable surgical scalpel, bandages, hospital beds, enema kits, manual stethoscope, bedpans, examination gloves, handheld surgical instruments, lab equipment analyzer, syringes, ear irrigation kit, infusion stand, medical disposable scissors, manual patient transfer device, urinal, bedpan, skin pressure protector, medical insoles, patient examination gloves, prep kit, hypodermic needle, pressure infuser for IV bags, and patient scale.

Medical-grade plastics and elastomers that have met FDA's biocompatibility and sterilization requirements are used for part manufacture. These materials are subdivided into categories based on the material performance:

(a) Low-performance polymers
(b) Medium-performance polymers
(c) High-performance polymers

4.2.1 Low-performance polymers

Low-performance polymers are typically commodity polymers that include the following:

(a) Polypropylene (PP)
(b) Polyethylene (PE)
(c) Polyvinyl chloride (PVC)
(d) Polystyrene (PS)

4.2.1.1 Polypropylene
PP is a semicrystalline material that is produced by the catalytic polymerization of propene. PPs are standard medical-grade plastics. Certain medical grades of PPs are stable at high temperatures. The following are the properties of PP:
- Excellent chemical resistance
- High purity
- Low water absorption
- Good electrical insulating properties
- Excellent dimensional stability
- Machinability
- Compatible with different methods of sterilization

4.2.1.2 Polyethylene
PE is a semicrystalline material that is produced by the addition or radical polymerization of ethylene (olefin) monomers in the presence of either Ziegler–Natta or Metallocene catalysts.

PE is a standard medical-grade plastic. The following are the properties:
- Excellent chemical and impact resistance
- Low coefficient of friction
- Resistant to major solvents
- Good fatigue and wear resistance
- Zero water absorption

4.2.1.3 Polyvinyl chloride
PVC is a thermoplastic polymer which is produced by free radical polymerization of vinyl chloride. PVC is extensively used in medical application due to following properties:
- Compatible with different sterilization methods, i.e., steam, radiation, and EtO
- Good flexibility
- Resistance to tears, scratches, and kinks
- Excellent insulation properties
- Good dimensional stability
- Resistant to chemicals and corrosion
- Low water absorption

4.2.1.4 Polystyrene

PS is a synthetic aromatic polymer that is produced when styrene undergoes polymerization in the presence of either heat or initiators. Benzoyl peroxide and di-tert-butyl per-benzoate are typical initiators used in the polymerization process. Active free radicals are formed when initiator thermally decomposes, thereby starting the polymerization process.

PS is used in medical application due to following properties:
- Superior clarity
- Good dimensional stability
- Excellent resistance to gamma radiation
- Poor chemical resistance to organic reagents
- Good dielectric strength

4.2.2 Medium-performance polymers

Medium-performance polymers are typically engineering polymers that include the following:
(a) Polycarbonate (PC)
(b) Polymethyl methacrylate (PMMA)
(c) Polybutylene terephthalate (PBT)
(d) Polyphenylene oxide (PPO)
(e) Acrylonitrile butadiene styrene (ABS)

4.2.2.1 Polycarbonate

PC is produced as a result of condensation polymerization reaction of bis-phenol A (produced through the condensation of phenol with acetone under acidic conditions) with carbonyl chloride in an interfacial process. Bis-phenol A is produced through the condensation of phenol with acetone in acidic conditions. PC is an amorphous polymer with good electrical properties combined together with superior impact strength, toughness, and moderate chemical resistance.

PC is an amorphous polymer that offers very high clarity. Other key properties include the following:
- Good electrical properties
- Superior impact strength and toughness
- Moderate chemical resistance
- High heat distortion temperature
- High dimensional stability
- Biocompatible

4.2.2.2 Polymethyl methacrylate
PMMA is produced by free-radical polymerization of methyl methacrylate (MMA).

PMMA is a transparent plastic that has high impact strength, shatter resistant, and excellent scratch resistance. Other key properties of PMMA that makes it material of choice for medical device and diagnostic equipment include the following:
- Biocompatibility
- Resistant to chemicals
- Dimensionally stability
- Optical clarity

4.2.2.3 Polybutylene terephthalate
PBT is a semicrystalline polymer produced by the polymerization of butanediol and terephthalic acid.

PBT has many attractive properties such as high strength and toughness, good abrasion, and heat resistance. In addition, there are other key properties such as follows:
- Low creep at elevated temperatures
- Good chemical resistance
- Excellent dimensional stability
- High rigidity
- Low water absorption
- Excellent heat aging

4.2.2.4 Polyphenylene oxide
Polyphenylene Oxide (PPO) or also referred to as polyphenylene ether is a crystalline polymer that is produced by oxidative coupling reaction between 2,6-dimethylphenol (DMP) and oxygen.

PPO has an outstanding mechanical, thermal, and electrical properties and can be sterilized. Other additional properties include the following:
- Excellent heat distortion resistance but unable to be melt processed. Therefore, it is used by blending with PS
- Very low moisture absorption
- Low thermal expansion
- Good dimensional stability

4.2.2.5 Acrylonitrile butadiene styrene

ABS is a terpolymer made by polymerizing styrene and acrylonitrile in the presence of polybutadiene. The proportions can vary from 15% to 35% acrylonitrile, 5% to 30% butadiene, and 40% to 60% styrene. The result is a long chain of polybutadiene crisscrossed with shorter chains of poly(styrene-co-acrylonitrile).

ABS is an amorphous copolymer with excellent impact, stiffness, and strength. In addition, ABS has following properties:
- Good chemical resistance
- Good dimensional stability
- Excellent abrasion resistance
- Biocompatible
- Compatible with gamma radiation and ethylene oxide (EtO) sterilization methods

4.2.3 High-performance polymers

High-performance polymers are typically specialty polymers that include the following:

(a) Acetal copolymer (POM-C)
(b) Polyether ether ketone (PEEK)
(c) Polyphenyl sulfone (PPSU)
(d) Polysulfone (PSU)
(e) Polyphenyl sulfide (PPS)
(f) Polyvinylidene fluoride (PVDF)
(g) Polyetherimide (PEI)
(h) Polydimethylsiloxane (PDMS)
(i) Thermoplastic polyurethane (TPU)
(j) Thermoplastic elastomer (TPE)

4.2.3.1 Acetal copolymer

POM-C is a strong, rigid, and nonporous engineering thermoplastic. It has low coefficient of friction and can be submerged in water. It is resistant to dilute acids, many cleaning agents, and solvents. It has very low moisture absorption and can be sterilized. POM-C is not for internal use in the body but used for applications that require ongoing sterilization. Examples include the following:
- Instrument handles
- Prosthesis part testing
- Trays

4.2.3.2 Polyether ether ketone

PEEK is a semicrystalline, engineering thermoplastic offering excellent chemical compatibility, low susceptibility to stress cracking, good electrical insulation, excellent mechanical strength and impact properties, and machinability. PEEK has good dimensional stability, is biocompatible, and can be sterilized using steam autoclave, EtO, and gamma radiation. Some of the applications of PEEK include the following:

- Dental instruments
- Surgical grips
- Targeting devices
- Endoscopic equipment
- Analytical and diagnostic equipment
- Biotechnology and laboratory equipment

4.2.3.3 Polyphenyl sulfone

PPSU is an amorphous material, with high glass transition temperature and low moisture absorption. PPSU offers exceptional hydrolytic stability, toughness, and superior impact strength over a wide temperature range. It also offers high deflection temperatures and outstanding resistance to environmental stress cracking. It can be sterilized by steam autoclaving, dry heat, EtO, and gamma radiation. PPSU is commonly used in the following:

- Sterilization trays
- Dental and surgical instrument handles
- Fluid handling coupling and fittings

4.2.3.4 Polysulfone

PSU is a high-temperature thermoplastic with high mechanical strength and rigidity. It offers good creep strength over a wide range of temperatures. PSU also has excellent dimensional stability, very good resistance to hydrolysis, and good chemical compatibility. PSU has good mechanical strength, resistance to sterilization, and stress cracking. Some examples include the following:

- Surgical trays
- Nebulizers
- Humidifiers

4.2.3.5 Polyphenylene sulfide

Polyphenyl sulfide (PPS) is a semicrystalline, high-temperature thermoplastic polymer. It has good resistance to chemicals, excellent electrical resistance, and exceptional mechanical strength, even at temperatures above 200°C. It has very good dimensional stability, low water absorption, and low susceptibility to creep. It is used in surgical instruments that require high dimensional stability, strength, and heat resistance. Examples include the following.

4.2.3.6 Polyvinylidene fluoride
PVDF is an opaque, semicrystalline thermoplastic fluoropolymer. It offers excellent chemical resistance, excellent mechanical strength, high dielectric strength, abrasion resistance, creep resistance, and low moisture absorption. It is compatible with gamma sterilization. PVDF is often used:
- Coating and legibly marking identification material to reusable and disposable medical devices and instruments due to its rugged wearability, chemical resistance, and inert characteristics
- As an ultrafine fiber filtration media for high-efficiency face masks
- In minimally invasive catheters

4.2.3.7 Polyetherimide
PEI material is an amorphous polymer, and its properties include high mechanical strength and rigidity at elevated temperatures, long-term heat resistance, dimensional stability, and good electrical properties. It also has a remarkably high creep resistance over a wide range of temperatures. and a high permanent operating temperature. It is an inherent flame retardant, resist hydrolysis, and compatible with steam, EtO, and gamma radiation sterilization. It is often used in the following:
- Chemical lab equipment and pharmaceutical equipment manifolds
- Containers and accessories
- Medical monitor probe housing
- Surgical instrument handles and enclosures
- Nonimplant prosthesis

4.2.3.8 Polydimethylsiloxane
Polydimethylsiloxane or PDMS is a silicone elastomer. It is biocompatible and chemically inert and provides a surface that has a low interfacial free energy. It also has good gas permeability and good thermal stability and is optically transparent. Some of PDMS examples include the following:
- Catheter and drainage tubing
- Dialysis membrane

- Micropumps
- Microvalves
- Adaptive lenses

4.2.3.9 Thermoplastic polyurethane
TPU is a thermoplastic elastomer offering high durability and flexibility. Key properties include abrasion and scratch resistance, UV resistance, and high transparency. Because of its excellent mechanical properties, durability, and resistance against oils and chemicals, examples of medical TPU applications include the following:
- Devices used for diagnostic, anesthesia, and artificial respiration
- Healthcare mattresses
- Dental materials
- Compression stockings
- Medical instrument cables
- Gel shoe orthotics
- Wound dressings

4.2.3.10 Thermoplastic elastomer
TPE is a thermoplastic that offers combined properties of plastic and elastomer. TPE is safe and biocompatible and has irritation-free properties. It has excellent barrier properties and sterilizable by steam, EtO, and gamma radiation. Some of the examples include the following:
- Medical tubes
- Syringe baskets
- Uterine balloon tamponades
- Oxygen masks
- Surgical elastomer wipes
- Disposable surgical gowns

References
[1] Medical Device Classification and Plastic Material Selection, Solvay. https://www.solvay.com/en/chemical-categories/specialty-polymers/healthcare/medical-device-classification-and-plastic.
[2] HCL Technologies, An Overview of the Plastics Material Selection for Medical Device, February 2013, https://www.hcltech.com/white-papers/engineering-rd-services/overview-plastic-material-selection-process-medical-devices
[3] Food and Drug Administration, Biocompatibility Evaluation Endpoints. https://www.fda.gov/medical-devices/biocompatibility-assessment-resource-center/biocompatibility-evaluation-endpoints-device-category.

[4] Food and Drug Administration, Appendix A, Evaluation Endpoints for Consideration, Use of International Standard ISO 10993-1, "Biological Evaluation of Medical Devices — Part 1: Evaluation and Testing within a Risk Management Process" — Guidance for Industry and Food and Drug Administration Staff, September 2020.
[5] M. Pathmajeyan, Cleaning, Disinfection and Sterilization Design Consideration for Medical Cables, Clearpath Medical, March 2020. https://clearpathmedical.com/cleaning-disinfections-and-sterilization-design-considerations-for-medical-cables/.
[6] R. Fenton, What are the Differences in the FDA Medical Device Classes, Qualio Blog: Quality & Compliance Hub, June 2021. https://www.qualio.com/blog/fda-medical-device-classes-differences.

CHAPTER FIVE

Low performance demand plastics and elastomers for medical devices

Contents

5.1 Introduction	79
5.2 Polypropylene	80
5.2.1 Properties	80
5.2.2 Applications	81
5.2.3 Medical grades and suppliers	82
5.3 Polyethylene	82
5.3.1 Properties	82
5.3.2 Applications	83
5.3.3 Medical grades and suppliers	84
5.4 Polyvinyl chloride (PVC)	85
5.4.1 Properties	85
5.4.2 Applications	86
5.4.3 Medical grades and suppliers	86
5.5 Polystyrene	87
5.5.1 Properties	87
5.5.2 Applications	88
5.5.3 Medical grades and suppliers	88
References	89

5.1 Introduction

Low-performance demand polymers (plastics and elastomers) are commodity polymers that are known to have desirable mechanical, thermal, and chemical properties. However, performance and cost of these polymers are less than specialty and engineering polymers, while production volumes are typically higher (see Fig. 5.1).

Low-performance demand plastics and elastomers are commodity polymers, and only the following selected polymers will be reviewed:

(a) Polypropylene (PP)
(b) Polyethylene (PE)

(c) Polyvinyl chloride (PVC)
(d) Polystyrene (PS)

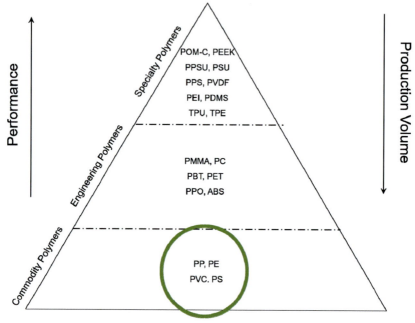

Figure 5.1 Performance demand tree for plastics and elastomers.

5.2 Polypropylene

5.2.1 Properties

PP is a semicrystalline material that is produced by the catalytic polymerization of propene. The structure of PP is shown in Fig. 5.2 [1].

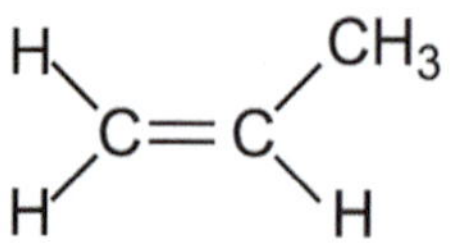

$$\underset{H}{\overset{H}{>}}C=C\underset{H}{\overset{CH_3}{<}}$$

Figure 5.2 Structure of polypropylene (PP) [1].

Polypropylenes are standard medical grade plastics. Certain medical grades of polypropylenes are stable at high temperatures. Table 5.1 shows physical, mechanical, and thermal properties of PP.

Additional properties of polypropylene include:
- Excellent chemical resistance
- High purity
- Low water absorption

- Good electrical insulating properties
- Excellent dimensional stability
- Machinability
- Compatible with different methods of sterilization

Table 5.1 Physical, mechanical, and thermal properties of polypropylene (PP).

Density (g/cm^3)	0.91
Water absorption, 24 h (%)	Slight
Tensile strength (psi)	9500
Flexural modulus (psi)	225,000
IZOD impact notched (ft-lb/in)	1.2
Coefficient of linear thermal expansion ($\times\ 10^{-5}$ in./in./°F)	5.0
Heat deflection temperature at 264 psi (°F)	210
Maximum continuous service temperature in air (°F)	180
Melting point (°F)	320

5.2.2 Applications

Excellent chemical resistance, dimensionally stability, and compatibility toward different sterilization method makes PP a material of choice for applications that requires steam sterilization. Fig. 5.3 shows an inhaler made from polypropylene) [2]. Additional examples include:
- Disposable syringes
- Inhalators
- Pumps
- Connectors
- Reusable plastic containers
- Dispensers
- Pharmacy prescription bottles
- IV bags

Figure 5.3 An inhaler made from polypropylene (PP) [2].

5.2.3 Medical grades and suppliers

Table 5.2 shows major producers of PP in the global medical market.

Table 5.2 Major suppliers of PP in the medical market.

Supplier	Medical grades
Braskem	Medcol
Exxon Mobil	PP1013H1, PP1014H1, PP9074MED
Ineos	Eltex MED
LyondellBasell	Purell HP548N, Purell RP315M, Purell EP374M, Purell GA7760, Purell HP371P, Purell HP372P, Purell HP373P, Purell HP570R, Purell RP373R, Purell RP374R, Purell RP375R

5.3 Polyethylene

5.3.1 Properties

PE is a semicrystalline material that is produced by the addition or radical polymerization of ethylene (olefin) monomers in the presence of either Ziegler–Natta or metallocene catalysts. Structure of PE is shown in Fig. 5.4 [3].

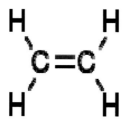

Figure 5.4 Structure of polyethylene (PE) [3].

Polyethylene (PE) is a standard medical grade plastic. Table 5.3 shows physical, mechanical, and thermal properties of PE.

Additional properties of polyethylene includes:
- Excellent chemical and impact resistance
- Low coefficient of friction

- Resistant to major solvents
- Good fatigue and wear resistance
- Zero water absorption

Table 5.3 Physical, mechanical, and thermal properties of polyethylene.

Physical properties	
Density (g/cm^3)	0.96
Water absorption, 24 h (%)	0.10
Mechanical properties	
Tensile strength (psi)	4000
Tensile elongation at break (%)	600
Flexural modulus (psi)	200,000
Thermal properties	
Coefficient of linear thermal expansion ($\times 10^{-5}$ in./in./°F)	7.0
Heat deflection temperature at 264 psi (°F)	172
Melting point (°F)	230

5.3.2 Applications

PE is a versatile and durable thermoplastic. Properties such as excellent impact resistance and resistance to chemicals, along with zero moisture absorption making it a preferred material for medical applications [4]. Some of the areas where PE is used include following:

- Sanitary packaging
- Labware
- Dental instruments
- Catheter tubing
- Implants

Fig. 5.5 shows application of PE. MEDPOR® porous polyethylene implants is produced by Stryker [5].

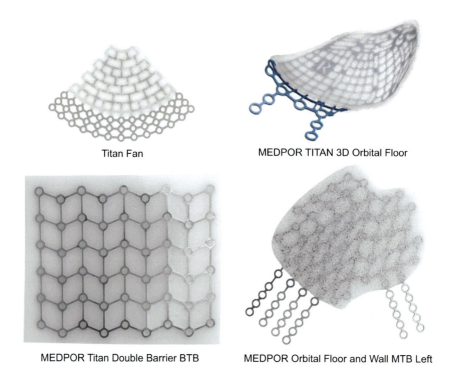

Titan Fan MEDPOR TITAN 3D Orbital Floor

MEDPOR Titan Double Barrier BTB MEDPOR Orbital Floor and Wall MTB Left

Source: Images are republished with permission from Stryker.

Figure 5.5 MEDPOR® Porous Polyethylene Implants [5].

5.3.3 Medical grades and suppliers

Some of the major suppliers in the PE medical market are listed in Table 5.4.

Table 5.4 Major suppliers in the polyethylene medical market.

Supplier	Medical grade
Dow	Health+
Exxon Mobil	Exceed XP, Exceed
LyndellBasell	Purell ACP 5531B, Purell ACP 6031D, Purell ACP 6541A, Purell GA 7760, Purell GB 7250, Purell GC 7260, Purell GC 7260G, Purell PE GF4750, Purell PE GF4760, Purell 2007H, Purell 2410T, Purell PE 1810E, Purell PE 1840H, Purell PE 2420F, Purell PE 3020D, Purell PE 3020K, Purell PE 3040D, Purell PE 3220D, Purell PE 3420F
SABIC	PCG02, PCG80, PCG01, PCG00, PCG06, PCG3054
Repsol Healthcare	HLD02G, HLD02S, HHD50G, HHD55G, HHD62G

5.4 Polyvinyl chloride (PVC)
5.4.1 Properties

PVC is a thermoplastic polymer which is produced by free radical polymerization of vinyl chloride. The structure of PVC is shown in Fig. 5.6 [6]:

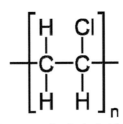

Figure 5.6 Structure of polyvinyl chloride (PVC) [6].

PVC is extensively used in medical application due to the following properties:
- Compatible with different sterilization methods, i.e., steam, radiation, and EtO
- Good flexibility
- Resistance to tears, scratches, and kinks
- Excellent insulation properties
- Good dimensional stability
- Resistant to chemicals and corrosion
- Low water absorption

Table 5.5 shows physical, mechanical, thermal, and electrical properties of PVC.

Table 5.5 Physical, mechanical, thermal, and electrical properties of polyvinyl chloride.

Physical properties	
Density (g/cm^3)	1.42
Water absorption, 24 h (%)	0.06
Mechanical properties	
Tensile strength (psi)	7500
Tensile modulus (psi)	411,000
Flexural strength (psi)	12,800

(*Continued*)

Table 5.5 Physical, mechanical, thermal, and electrical properties of polyvinyl chloride.—cont'd

Physical properties	
Flexural modulus (psi)	481,000
IZOD impact notched (ft-lb/in)	1.0
Thermal properties	
Heat deflection temperature at 264 psi (°F)	158
Melting point (°F)	185
Maximum operating temperature (°F)	140
Electrical properties	
Dielectric strength, V/mil	544

5.4.2 Applications

Owing to good dimensional stability, flexibility, resistance toward tears, scratches, and kinks, resistance to chemicals and corrosion, and compatibility with different sterilization methods, PVC is widely used in the following:
- Medical tubing
- Connectors
- Catheters
- Medical components
- Dental components
- Hospital flooring
- Wall covering
- Blood bags
- Bottles and jars

5.4.3 Medical grades and suppliers

Some of the major suppliers in the PVC medical market are listed in Table 5.6.

Table 5.6 Major supplier in the polyvinyl chloride medical market.

Supplier	Medical grade
Alphagary	Alphamed PVC Compounds
Formosa	Formolon Suspension PVC
Avient	CORE Vinyl Plastisols
Tekni-Plex	Flexchem, Flexchem SR

5.5 Polystyrene
5.5.1 Properties

PS is a synthetic aromatic polymer that is produced when styrene undergoes polymerization in the presence of either heat or initiators. Benzoyl peroxide and di-tert-butyl per-benzoate are typical initiators used in the polymerization process. Active free radicals are formed when initiator thermally decomposes, thereby starting the polymerization process. Structure of pure polystyrene is shown in Fig. 5.7 [7]:

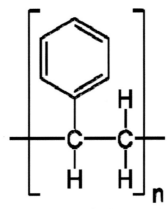

Figure 5.7 Structure of polystyrene (PS) [7].

PS is used in medical application due to the following properties:
- Superior clarity
- Good dimensional stability
- Excellent resistance to gamma radiation
- Poor chemical resistance to organic reagents
- Good dielectric strength

Table 5.7 shows physical, mechanical, thermal, and electrical properties of PS.

Table 5.7 Physical, mechanical, thermal, and electrical properties of polystyrene.

Physical properties

Density (g/cm^3)	1.04
Water absorption, 24 h (%)	0.3

Mechanical properties

Tensile strength (psi)	7700
Tensile elongation at break (%)	3.4
Flexural strength (psi)	120,000
Flexural modulus (psi)	350,000
IZOD impact notched (ft-lb/in)	1.2

Thermal properties

Heat deflection temperature at 264 psi (°F)	203
Melting point (°F)	518
Maximum operating temperature (°F)	122

Electrical properties

Dielectric strength, V/mil	600

5.5.2 Applications

With its low cost, superior clarity, dimensional stability, and adaptability to radiation sterilization, PS possesses many features for medical applications. Medical applications of general purpose polystyrene (GPPS) include labware such as follows:

- Tissue culture trays
- Test tubes
- Petri dishes

Both high-impact PS and general-purpose PS find uses in medical applications such as follows:

- Diagnostic components
- Housing for test kits
- Syringe hubs
- Respiratory care equipment
- Suction canisters

5.5.3 Medical grades and suppliers

There are several suppliers that produce non—medical-grade PS, however. As shown in Table 5.8, Trinseo Styron is the only resin supplier that supports PS medical market.

Table 5.8 Major resin supplier for the polystyrene medical market.

Supplier	Medical grade
Trinseo	STYRON 2678 MED

References

[1] Polypropylene, The Essential Chemical Industry. https://www.essentialchemicalindustry.org/polymers/polypropene.html.
[2] Symbicort Dose Inhaler, Wikimedia Commons. https://en.wikipedia.org/wiki/Inhaler#Medical_uses.
[3] Polyethylene, Wikipedia. https://en.wikipedia.org/wiki/Polyethylene.
[4] B. Poon, L. Czuba, in: M.A. Spalding, A.M. Chatterjee (Eds.), Chapter 44: Medical Applications of Polyethylene, Handbook of Industrial Polyethylene and Technology: Definitive Guide to Manufacturing, Properties, Processing, Applications and Market Set, Wiley Publishers, 2017, p. 1155.
[5] MEDPOR® Porous Polyethylene Implants, Stryker. https://www.strykerneurotechnology.com/medpor-porous-polyethylene-implants.
[6] Polyvinyl Chloride, Wikipedia. https://simple.wikipedia.org/wiki/Polyvinyl_chloride.
[7] Polystyrene, Wikipedia. https://en.wikipedia.org/wiki/Polystyrene.

CHAPTER SIX

Medium-performance demand plastics and elastomers for medical devices

Contents

6.1 Introduction	91
6.2 Polycarbonate	92
6.2.1 Properties	92
6.2.2 Applications	93
6.2.3 Medical grades and suppliers	94
6.3 Polymethyl methacrylate	95
6.3.1 Properties	95
6.3.2 Applications	96
6.3.3 Medical grades and suppliers	96
6.4 Polybutylene terephthalate	97
6.4.1 Properties	97
6.4.2 Applications	98
6.4.3 Medical grades and suppliers	98
6.5 Polyphenylene oxide	98
6.5.1 Properties	98
6.5.2 Applications	100
6.5.3 Medical grades and suppliers	100
6.6 Acrylonitrile butadiene styrene	100
6.6.1 Properties	100
6.6.2 Applications	102
6.6.3 Medical grades and suppliers	102
References	102

6.1 Introduction

Medium-performance demand polymers (plastics and elastomers) are engineering polymers that are known to have desirable mechanical, thermal, and chemical properties. As shown in Fig. 6.1, the performance, cost, and the production volumes of these polymers are in between commodity and specialty polymers.

Applications of Polymers and Plastics in Medical Devices
ISBN: 978-0-12-820980-6
https://doi.org/10.1016/B978-0-12-820980-6.00015-1
© 2022 Elsevier Inc.
All rights reserved.

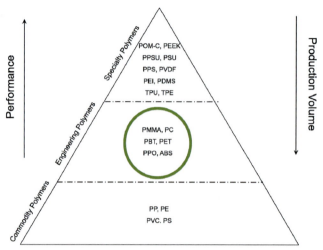

Figure 6.1 Performance demand tree for plastics and elastomers.

List of medium-performance demand plastics and elastomers includes the following:
(a) Polycarbonate (PC)
(b) Polymethyl methacrylate (PMMA)
(c) Polybutylene terephthalate (PBT)
(d) Polyphenylene oxide (PPO)
(e) Acrylonitrile butadiene styrene (ABS)

6.2 Polycarbonate

6.2.1 Properties

Polycarbonate is produced as a result of condensation polymerization reaction of bis-phenol A (produced through the condensation of phenol with acetone under acidic conditions) with carbonyl chloride in an interfacial process. Bis-phenol A is produced through the condensation of phenol with acetone in acidic conditions. The structure of polycarbonate is shown in Fig. 6.2 [1].

Figure 6.2 Structure of polycarbonate [1].

PC is an amorphous polymer that offers very high clarity. Other key properties include the following:
- Good electrical properties
- Superior impact strength and toughness
- Moderate chemical resistance
- High heat distortion temperature
- High-dimensional stability
- Biocompatible

Table 6.1 shows physical, mechanical, and thermal properties of PC.

Table 6.1 Physical, mechanical, and thermal properties of polycarbonate.

Density (g/cm^3)	1.22
Water absorption, 24 h (%)	0.15
Tensile strength (psi)	9500
Flexural modulus (psi)	345,000
IZOD impact notched (ft-lb/in)	16
Coefficient of linear thermal expansion ($\times 10^{-5}$ in./in./°F)	3.8
Heat deflection temperature at 264 psi (°F)	270
Maximum continuous service temperature in air (°F)	240
Melting point (°F)	265

6.2.2 Applications

PC is the material of choice in medical applications that require clarity, strength, heat resistance, biocompatibility, and low water absorption. PC is used in following life-supporting devices [2]:
- Hemodialyzers
- Anesthesia containers
- Blood oxygenators
- Arterial filters
- Intravenous connectors
- Endoscopic appliances

Pictorial representation of devices made from polycarbonate are shown in Fig. 6.3 [3,4].

Source: Image reproduced with permission from Covestro
Luer – Covestro's Makrolon Rx3440

Source: Image reproduced with permission from TERUMO Medical
CAPIOX Adult Arterial Filter – TERUMO Cardiovascular

Figure 6.3 Examples of devices made from polycarbonate (PC) [3,4].

6.2.3 Medical grades and suppliers

There are many players in the market that supply PC; however, Covestro and SABIC account for approximately half of the global demand. Table 6.2 shows medical grade(s) that are supplied by Covestro and SABIC for medical applications.

Table 6.2 Major suppliers of polycarbonate in the medical market.

Supplier	Medical grades
Covestro	Makrolon
SABIC	LEXAN

6.3 Polymethyl methacrylate

6.3.1 Properties

PMMA is produced by free-radical polymerization of methyl methacrylate (MMA) as shown in Fig. 6.4 [5].

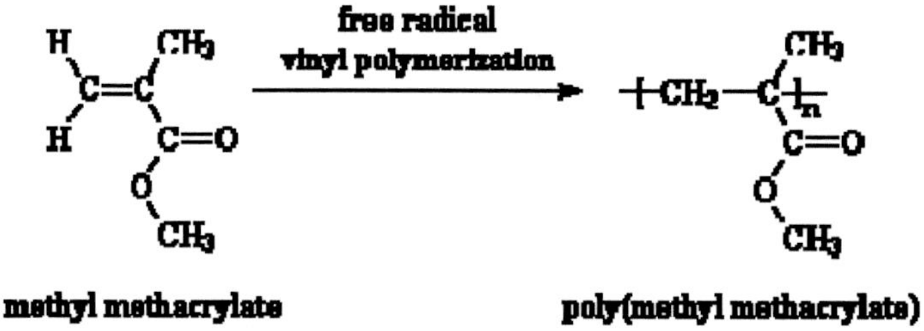

Figure 6.4 Structure of polymethyl methacrylate (PMMA) [6].

PMMA is a transparent plastic that has high impact strength, shatter resistant, and excellent scratch resistance. Other key properties of PMMA that makes it material of choice for medical device and diagnostic equipment include the following:
- Biocompatibility
- Resistant to chemicals
- Dimensional stability
- Optical clarity

Table 6.3 shows physical, mechanical, and thermal properties of PMMA.

Table 6.3 Physical, mechanical and thermal properties of polymethyl methacrylate.

Physical properties	
Density (g/cm^3)	1.19
Water absorption, 24 h (%)	0.20
Mechanical properties	
Tensile strength (psi)	10,000
Tensile modulus (psi)	400,000
Tensile elongation at break (%)	4.5
Flexural strength (psi)	17,000
Flexural modulus (psi)	480,000
IZOD impact notched (ft-lb/in)	0.4
Hardness	M95
Thermal properties	
Coefficient of linear thermal expansion ($\times\ 10^{-5}$ in./in./°F)	4.0
Heat deflection temperature at 264 psi (°F)	195
Melting point (°F)	320
Maximum operating temperature (°F)	160
Dielectric strength (V/mil)	430

6.3.2 Applications

PMMA is the material of choice for medical applications where transparency, dimensional stability, chemical resistance, and biocompatibility are needed. PMMA is used for a range of applications in medical device and labware applications.

Fig. 6.5 shows EVONIK's Cyrolite® Protect 2 with built-in antimicrobial agents made from PMMA [7,8].

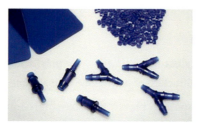

Figure 6.5 Evonik's CYROLITE® Protect 2 with built-in antimicrobial agents made from PMMA [7]. *Source: Image reproduced with permission from EVONIK.*

Some other examples include:
- Dialysis and blood therapy systems
- IV components and drug delivery
- Devices
- Luers
- Connectors
- Spikes
- Filter housing
- Respiratory canisters
- In vitro diagnostic systems

6.3.3 Medical grades and suppliers

Some of the major suppliers in the PMMA medical market are listed in Table 6.4.

Table 6.4 Major suppliers in the polymethyl methacrylate (PMMA) medical market.

Supplier	Medical grade
Arkema	ALTUGLAS
Evonik	CYROLITE

6.4 Polybutylene terephthalate
6.4.1 Properties

Polybutylene Terephthalate (PBT) is a semicrystalline polymer produced by the polymerization of butanediol and terephthalic acid as shown in Fig. 6.6 [8].

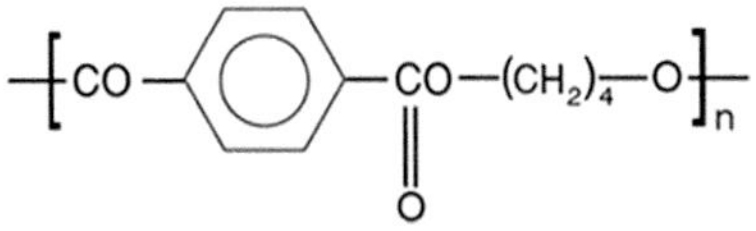

Figure 6.6 Structure of polybutylene terephthalate (PBT) [8].

PBT has many attractive properties such as high strength and toughness, good abrasion, and heat resistance. Table 6.5 shows physical, mechanical, thermal, and electrical properties of PBT.

Table 6.5 Physical, mechanical, thermal, and electrical properties of polybutylene terephthalate.

Physical properties	
Density (g/cm^3)	1.30
Water absorption, 24 h (%)	0.08
Mechanical properties	
Tensile strength (psi)	8690
Tensile modulus (psi)	416,000
Tensile elongation at break (%)	300
Flexural strength (psi)	12,000
Flexural modulus (psi)	330,000
IZOD impact notched (ft-lb/in)	1.5
Hardness	M72
Thermal properties	
Heat deflection temperature at 264 psi (°F)	310
Melting point (°F)	433
Maximum operating temperature (°F)	245
Electrical properties	
Dielectric strength, V/mil	400

Other key properties of PBT includes:
- Low creep at elevated temperatures
- Good chemical resistance
- Excellent dimensional stability
- High rigidity
- Low water absorption
- Excellent heat aging

6.4.2 Applications

PBT offers high rigidity and strength, excellent dimensional stability, low water absorption, good resistance to chemicals, and excellent heat aging. Because of these properties, PBT is used in following medical devices:
- Injection-molded connectors for pulse oximeters
- Tips for electrosurgical instruments
- Clips for breathing mask

6.4.3 Medical grades and suppliers

Some of the major suppliers in the PBT market are listed in Table 6.6.

Table 6.6 Major supplier in the polybutylene terephthalate market.

Supplier	Medical grade
BASF	Ultradur PRO
Celanese Corporation	Celanex MT PBT

6.5 Polyphenylene oxide

6.5.1 Properties

Polyphenylene Oxide (PPO) or also referred to as polyphenylene ether as shown in Fig. 6.7 [9]. It is a crystalline polymer that is produced by oxidative coupling reaction between 2,6-dimethylphenol (DMP) and oxygen as shown in Fig. 6.8 [10].

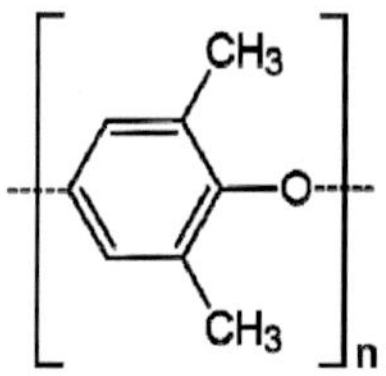

Figure 6.7 Structure of polyphenylene oxide (PPO) [9].

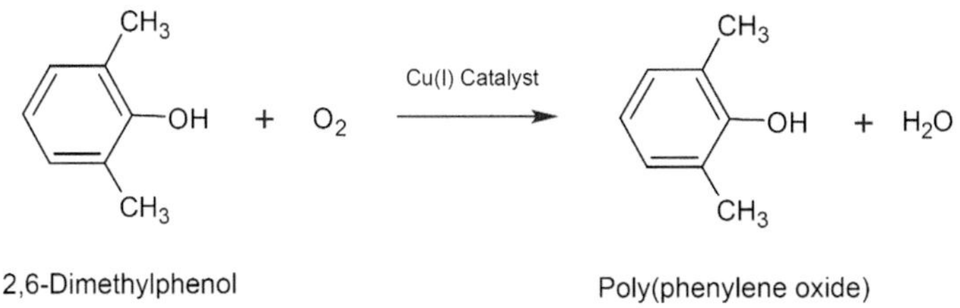

Figure 6.8 Synthesis of polyphenylene oxide (PPO) [10].

PPO has outstanding mechanical, thermal, and electrical properties and can be sterilized. Table 6.7 shows physical, mechanical, thermal, and electrical properties of PPO.

Table 6.7 Physical, mechanical, thermal, and electrical properties of polyphenylene oxide.

Property	Value
Physical properties	
Density (g/cm^3)	1.08
Water absorption, 24 h (%)	0.07
Mechanical properties	
Tensile strength (psi)	9600
Tensile modulus (psi)	350,000
Tensile elongation at break (%)	25
Flexural strength (psi)	14,400
Flexural modulus (psi)	370,000
IZOD impact notched (ft-lb/in)	3.5
Hardness	R119
Thermal properties	
Coefficient of linear thermal expansion ($\times 10^{-5}$ in./in./°F)	3.3
Heat deflection temperature at 264 psi (°F)	254
Melting point (°F)	Cannot be melted
Maximum operating temperature (°F)	220
Electrical properties	
Dielectric strength, V/mil	500

Other properties include:
- Excellent heat distortion resistance but unable to be melt processed. Therefore, it is used by blending with polystyrene
- Very low moisture absorption
- Low thermal expansion
- Good dimensional stability

6.5.2 Applications

PPO is used in applications where high heat resistance, dimensional stability, and accuracy are needed. Some examples include the following:
- Drug delivery inhaler
- Disposable dental X-ray mouth guard
- Surgical stapler components
- Sterilization trays
- Dental trays
- Surgical instrument handle

6.5.3 Medical grades and suppliers

Some of the major suppliers in the PPO medical market are listed in Table 6.8.

Table 6.8 Major supplier in the polyphenylene oxide medical market.

Supplier	Medical grade
SABIC	NORYL—HN731A, HN731SE, HNA033, HNA055

6.6 Acrylonitrile butadiene styrene

6.6.1 Properties

ABS is a terpolymer made by polymerizing styrene and acrylonitrile in the presence of polybutadiene. The proportions can vary from 15% to 35% acrylonitrile, 5%—30% butadiene and 40%—60% styrene. The result is a long chain of polybutadiene crisscrossed with shorter chains of poly(styrene-co-acrylonitrile) as shown in Fig. 6.9 [11].

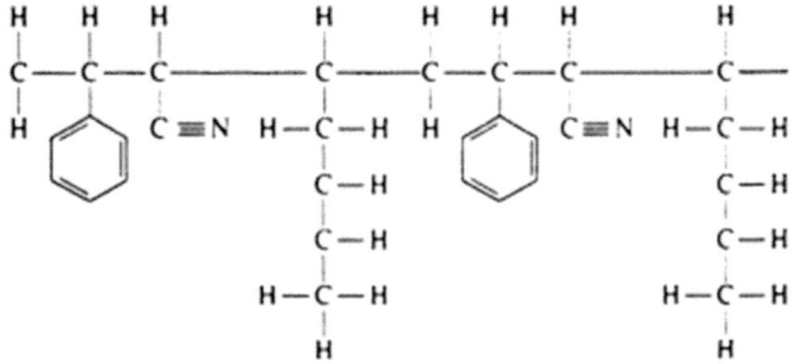

Medium-performance demand plastics and elastomers for medical devices

Figure 6.9 Structure of acrylonitrile butadiene styrene (ABS) [11].

ABS is an amorphous copolymer with excellent impact, stiffness, and strength. Table 6.9 shows physical, mechanical, thermal, and electrical properties of ABS.

Table 6.9 Physical, mechanical, and thermal properties of acrylonitrile butadiene styrene.

Physical properties	
Density (g/cm^3)	1.04
Water absorption, 24 h (%)	0.30
Mechanical properties	
Tensile strength (psi)	4100
Tensile modulus (psi)	294,000
Tensile elongation at break (%)	32
Flexural strength (psi)	9100
Flexural modulus (psi)	304,000
IZOD impact notched (ft-lb/in)	7.7
Hardness	R102
Thermal properties	
Coefficient of linear thermal expansion ($\times\ 10^{-5}$ in./in./°F)	5.6
Heat deflection temperature at 264 psi (°F)	177
Melting point (°F)	392
Maximum operating temperature (°F)	160

In addition, ABS has following properties [12]:
- Good chemical resistance
- Good dimensional stability
- Excellent abrasion resistance
- Biocompatible
- Compatible with gamma radiation and ethylene oxide (EtO) sterilization methods

6.6.2 Applications

ABS is used in applications that require product to be dimensionally stable, be able to withstand impact and exposure to chemicals, and be compatible with majority of sterilization methods.

Some examples where ABS is used includes:
- Dialyzer
- Surgical instruments
- IV three-way stopcock
- Inhaler cartridge
- Drug delivery syringe

6.6.3 Medical grades and suppliers

Some of the major suppliers in the ABS medical market are listed in Table 6.10.

Table 6.10 Major supplier in the acrylonitrile butadiene styrene medical market.

Supplier	Medical grades
Chi Mei	Polylac ABS
SABIC	Cycolac ABS—HMG47MD, HMG94MD
BASF	Terluran HD-15
Lanxess Cheil	Lustran ABS 348
Formosa Plastics	Tairilac
Trinseo	Magnum 3404
Styrolution	Novodur HD

References

[1] Polycarbonate, Wikipedia. https://en.wikipedia.org/wiki/Polycarbonate.
[2] D.E. Powell, Medical Applications of Polycarbonate, Medical Plastics and Biomaterials Magazine, Medical Device and Diagnostic. https://www.mddionline.com/news/medical-applications-polycarbonate.
[3] IV Connectors, Covestro. https://www.covestro.us/en/media/news-releases/2018-news-releases/2018-02-06-better-performing-iv-connectors.

[4] CAPIOX® Arterial Filter, Terumo Cardiovascular. https://www.terumo-europe.com/en-emea/products/capiox%C2%AE-arterial-filter-filter.
[5] Polymethylmethacrylate, Polymer Science Learning Center. https://pslc.ws/macrog/pmma.htm.
[6] Evonik launches CYROLITE Protect 2 for Antimicrobial Medical Devices, Med Device Online. https://www.meddeviceonline.com/doc/evonik-launches-cyrolite-protect-for-antimicrobial-medical-devices-0001.
[7] Complete Guide on Polybutylene Terephthalate, Omnexus. https://omnexus.specialchem.com/selection-guide/polybutylene-terephthalate-pbt-plastic.
[8] Poly(p-phenylene oxide), Wikipedia. https://en.wikipedia.org/wiki/Poly(p-phenylene_oxide.
[9] Polyphenylene Oxide, Polymer Properties Database. https://polymerdatabase.com/Polymer%20Brands/PPO.html.
[10] Acrylonitrile Butadiene Styrene, Selection Guide, Omnexus. https://omnexus.specialchem.com/selection-guide/acrylonitrile-butadiene-styrene-abs-plastic.
[11] Acrylonitrile Butadiene Styrene, Wikipedia. https://en.wikipedia.org/wiki/Acrylonitrile_butadiene_styrene#:~:text=ABS%20is%20a%20terpolymer%20made,styrene%2Dco%2Dacrylonitrile.

High-performance demand plastics and elastomers for medical devices

Contents

7.1 Introduction	106
7.2 Acetal copolymer	107
7.2.1 Properties	107
7.2.2 Applications	109
7.2.3 Medical grades and suppliers	109
7.3 Polyetheretherketone	109
7.3.1 Properties	109
7.3.2 Applications	110
7.3.3 Medical grades and suppliers	111
7.4 Polyphenyl Sulfone	112
7.4.1 Properties	112
7.4.2 Applications	113
7.4.3 Medical grades and suppliers	113
7.5 Polysulfone	113
7.5.1 Properties	113
7.5.2 Applications	115
7.5.3 Medical grades and suppliers	115
7.6 Polyphenylene sulfide	115
7.6.1 Properties	115
7.6.2 Applications	116
7.6.3 Medical grades and suppliers	117
7.7 Polyvinylidene fluoride	117
7.7.1 Properties	117
7.7.2 Applications	119
7.7.3 Medical grades and suppliers	119
7.8 Polyetherimide	119
7.8.1 Properties	119
7.8.2 Applications	120
7.8.3 Medical grades and suppliers	121
7.9 Polydimethylsiloxane	121
7.9.1 Properties	121
7.9.2 Applications	122
7.9.3 Medical grades and suppliers	122
7.10 Thermoplastic polyurethane	122

Applications of Polymers and Plastics in Medical Devices
ISBN: 978-0-12-820980-6
https://doi.org/10.1016/B978-0-12-820980-6.00014-X

© 2022 Elsevier Inc.
All rights reserved.

7.10.1 Properties	122
7.10.2 Applications	123
7.10.3 Medical grades and suppliers	124
7.11 Thermoplastic elastomer	124
7.11.1 Properties	124
7.11.2 Applications	125
7.11.3 Medical grades and suppliers	125
References	126

7.1 Introduction

High-performance demand polymers (plastics and elastomers) are a group of polymer materials that are known to retain their desirable mechanical, thermal, and chemical properties when subjected to harsh environment such as high temperature and pressure and corrosive chemicals. As shown in Fig. 7.1, these polymers excel in their performance when compared with commodity and engineering polymers. High-demand polymers are limited to specialty applications and generate low production volume.

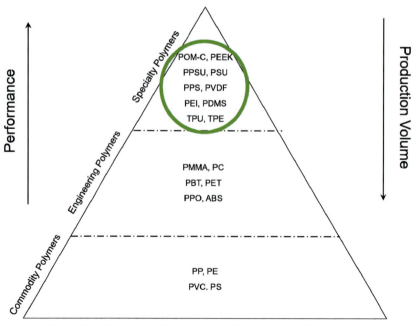

Figure 7.1 Demand performance tree for plastics and elastomers.

List of high-performance demand plastics and elastomers include the following:
(a) Acetal copolymer (POM-C)
(b) Polyether ether ketone (PEEK)
(c) Polyphenyl sulfone (PPSU)
(d) Polysulfone (PSU)
(e) Polyphenyl sulfide (PPS)
(f) Polyvinylidene fluoride (PVDF)
(g) Polyetherimide (PEI)
(h) Polydimethylsiloxane (PDMS)
(i) Thermoplastic polyurethane (TPU)
(j) Thermoplastic elastomer (TPE)

7.2 Acetal copolymer

7.2.1 Properties

Acetal copolymer also known as polyacetal and chemically referred to as "polyoxymethylene" (POM) is a functional group or molecule containing the functional group in which carbon C is bonded with two −OR groups as shown in Fig. 7.2 [1].

The acetal polymer molecules have a shorter backbone bond and packed together closely. This results in polymer being harder with high melting point. It is a highly crystalline thermoplastic that is known for its high flexural and tensile strength, stiffness, hardness, and low creep under stress. It also has a low coefficient of friction, excellent chemical resistance, and outstanding fatigue properties, but only moderate heat stability and insufficient flame resistance.

Acetal copolymer is produced by the ring-opening copolymerization of trioxane (the cyclic trimer of formaldehyde) with or without a small amount of cyclic ether (typically ethylene oxide or 1,3-dioxolane) as shown in Fig. 7.3 [2].

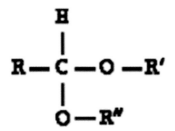

Figure 7.2 Structure of polyacetal [1].

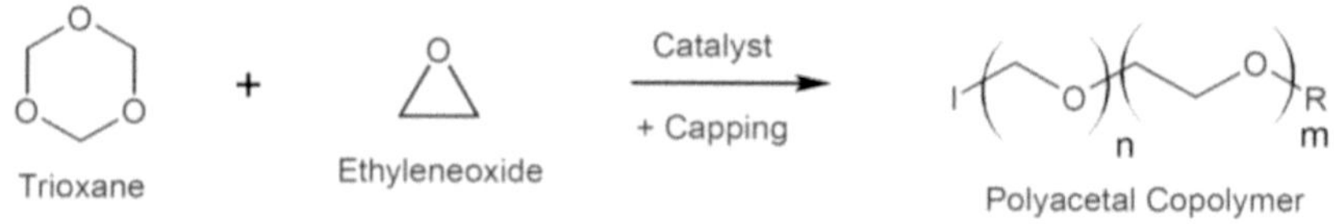

Figure 7.3 Ring opening copolymerization [2].

Key benefits of acetal copolymer include [3]:
- Excellent mechanical properties over a temperature range up to 140°C, down to −40°C
- High tensile strength, rigidity, and toughness (short-term)
- Low tendency to creep and fatigue (long-term). Not susceptible to environmental stress cracking
- High degree of crystallinity and excellent dimensional stability
- Excellent wear resistance
- Low coefficient of friction
- Good resistance to organic solvents and chemicals (except phenols) at room temperature
- Low smoke emission
- High gloss surfaces
- Low moisture absorption

Table 7.1 shows physical, mechanical, and thermal properties of acetal.

Table 7.1 Physical, mechanical, and thermal properties of acetal.

Physical properties	
Density (g/cm^3)	1.41
Water absorption, 24 h (%)	0.2
Mechanical properties	
Tensile strength (psi)	9500
Tensile modulus (psi)	400,000
Tensile elongation at break (%)	30
Flexural strength (psi)	12,000
Flexural modulus (psi)	400,000
Compressive strength (psi)	15,000
Compressive modulus (psi)	400,000
Hardness, Rockwell	M88/R120
IZOD impact notched (ft-lb/in)	1
Thermal properties	
Coefficient of linear thermal expansion ($\times\ 10^{-5}$ in./in./°F)	5.4
Heat deflection temperature at 264 psi (°F)	220
Melting point (°F)	335
Maximum operating temperature (°F)	180

7.2.2 Applications

Acetal copolymer is nonhygroscopic and is FDA complaint. It absorbs minimal amounts of moisture and can be cleaned easily. As a result, it finds applications in high moisture, food handling, medical, and marine components [4].

Acetal's other benefits include low water absorption, good electrical properties, resistance to fatigue and organic solvents, and outstanding wear characteristics [5].

Acetal copolymer is an excellent material for medical device applications that require low friction. They are also easy to machine, chemically resistant to hydrocarbons, neutral chemicals, and solvents. They make an excellent choice for medical device applications that require complex tight tolerances. Some of the medical applications include the following:

- Medical instruments that must be sterilized regularly, such as handles, trays, prosthesis parts, scraper blades
- MRI machines, surgical instruments, dental instruments, handheld diagnostic wands, and sterilization trays
- Endoscopic probes, diagnostic, anesthetic, and imaging equipment

7.2.3 Medical grades and suppliers

Among number of resin suppliers, there are seven top key suppliers of acetal copolymers as shown in Table 7.2. Although Dow Dupont is not included in this list, they are one of the top suppliers of Delrin, an acetal homopolymer.

7.3 Polyetheretherketone

7.3.1 Properties

Polyetheretherketone (PEEK) is a semicrystalline high-performance polymer that is produced using step-growth polymerization by the dialkylation of bisphenolate salts [6,7].

Table 7.2 Key suppliers and their commercially available medical grades.

Supplier	Medical grades
Westlake Plastics	Pomalux
Celanese	Celcon MT24U01
Modern Plastics	Tecaform AH MT
Quadrant	Acetron GP
BASF	Ultraform
ZL Engineering Plastics	ZL 900 Series
RadiciGroup	Heraform

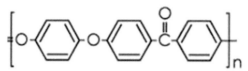

Figure 7.4 Structure of polyetheretherketone [8].

Structure of PEEK as shown in Figure 7.4 contains stiff aromatic polymer backbone which contributes towards higher thermal transitions. This results in polymer having continuous use temperature of around 240°C [8]. PEEK has excellent mechanical properties and outstanding chemical resistance, equivalent to fluoropolymers. The presence of either linkage provides improved melt processability. The rigid and stiff chemical structure affects the crystallinity of PEEK and allows up to 48% maximum achievable crystallinity [9].

Key properties of PEEK include the following [10]:
1. Low friction
2. Good dimensional stability
3. Exception insulation properties
4. Excellent sterilization resistance at high temperature
5. Biocompatible
6. Long life
7. Inherent purity
8. Excellent chemical compatibility
9. Low susceptibility to stress cracking
10. Excellent mechanical strength and impact properties
11. Machinability
12. Dimensionally stable

Table 7.3 shows physical, mechanical, and thermal properties of PEEK.

7.3.2 Applications

PEEK is the material of choice when it comes to spinal implant and trauma fixation [11].

Some other application areas include:
- Orthopedics and radiolucent implants
- Spine
- Sport medicine
- Cardiovascular
- Long term human implants
- As a replacement to metals in components for housing and surgical instruments

Table 7.3 Physical, mechanical, and thermal properties of PEEK.

Physical properties	
Density (g/cm^3)	1.32
Water absorption, 24 h (%)	0.5

Mechanical properties	
Tensile strength (psi)	14,000
Tensile modulus (psi)	490,000
Tensile elongation at break (%)	60
Flexural strength (psi)	24,600
Flexural modulus (psi)	590,000
IZOD impact notched (ft-lb/in)	1.6

Thermal properties	
Coefficient of linear thermal expansion ($\times\ 10^{-5}$ in./in./°F)	2.6
Heat deflection temperature at 264 psi (°F)	306
Melting point (°F)	649
Maximum operating temperature (°F)	338

7.3.3 Medical grades and suppliers

Some of the major suppliers in the PEEK market are listed in Table 7.4.

Table 7.4 Major suppliers in the polyetheretherketone market.

Supplier	Medical grade
Ensinger	Tecapeek
Mitsubishi	Ketron, Semitron
Avient	AUROtec, Edgetek, Lubri-Tech, Stat-Tech
Evonik	VESTAKEEP, VESTAPE
Lubrizol	Carbo-Rite
Rochling	SUSTAPEEK
Sabic	LNP LUBRICOMP, LNP STAT-KON, LNP THERMOCOMP
Saint-Gobain Performance Materials	Meldin
Solvay and Solvay Specialty Polymers	APC-2/AS4, APC-2/IM7, APC-2-PEEK, Ajedium, AvaSpire, KetaSpire, ZENIVA
Westlake Plastics	Arolux, MediPEEK

7.4 Polyphenyl Sulfone

7.4.1 Properties

Polyphenylene Sulfone (PPSU) is an amorphous material, with high glass transition temperature and low moisture absorption. It primarily consists of aromatic rings that are linked by sulfone groups as shown in Fig. 7.5 [12].

Table 7.5 shows physical, mechanical, thermal, and electrical properties of PPSU.

Figure 7.5 Structure of polyphenyl sulfone (PPSU) [12].

Table 7.5 Physical, mechanical, thermal, and electrical properties of polyphenyl sulfone.

Physical properties	
Density (g/cm^3)	1.29
Water absorption, 24 h (%)	0.37
Mechanical properties	
Tensile strength (psi)	10,100
Tensile modulus (psi)	340,000
Tensile elongation at break (%)	60–120
Flexural strength (psi)	13,200
Flexural modulus (psi)	350,000
IZOD impact notched (ft-lb/in)	13
Hardness	M80, R120, Shore D80
Thermal properties	
Coefficient of linear thermal expansion ($\times\ 10^{-5}$ in./in./°F)	3.1
Heat deflection temperature at 264 psi (°F)	405
Melting point (°F)	650–730
Maximum operating temperature (°F)	392
Electrical properties	
Dielectric strength, V/mil	360

PPSU offers exceptional hydrolytic stability, toughness, and superior impact strength over a wide temperature range. It also offers high deflection temperatures and outstanding resistance to environmental stress cracking. It can be sterilized by steam autoclaving, dry heat, EtO, and gamma radiation.

7.4.2 Applications

PPSU is the material of choice when an alternative to metal in sterilization cases and trays, implant trials, and surgical instrument handles is needed because of its superior impact resistance, exposure to high temperature, chemical disinfectants, hot water, and steam.

Other applications of PPSU include:
- Sterilization trays
- Dental and surgical instrument handles
- Fluid handling coupling and fittings
- Insulin pens, blood glucometers, tubing, sterilization trays, and cauterization devices
- End caps and other small components for pharmaceutical and electronics equipment
- Electronic assembly equipment
- Fluid handling couplings and fittings

7.4.3 Medical grades and suppliers

Some of the major suppliers in the PPSU market are listed in Table 7.6.

7.5 Polysulfone

7.5.1 Properties

Polysulfones (PSU) are prepared by polycondensation of suitable monomers or by ring-opening polymerization of cyclic ether sulfones. One of the most common methods is nucleophilic substitution of an aromatic chloro- or fluorosulfone by a phenoxide ion. PSU structure is shown in Fig. 7.6 [13].

Table 7.7 shows physical, mechanical, thermal, and electrical properties of PSU.

Table 7.6 Major supplier in the polyphenyl sulfone (PPSU) market.

Supplier	Medical grade
Solvay	Radel PPSU
Mitsubishi Chemical Advanced Materials	Sultron PPSU
UJU New Materials Co. Ltd.	Paryls PPSU

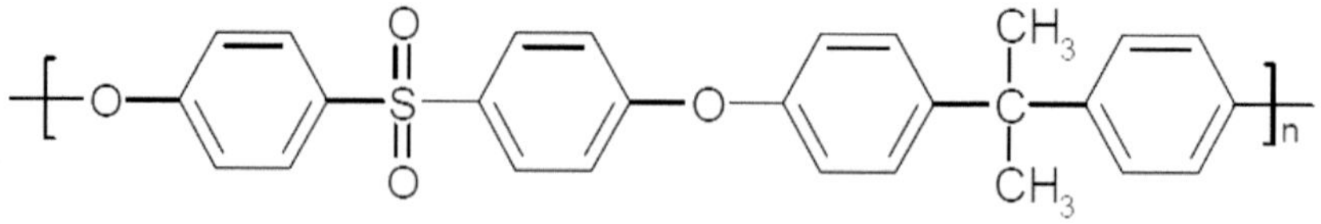

Figure 7.6 Structure of polysulfone (PSU) [13].

Table 7.7 Physical, mechanical, thermal, and electrical properties of polysulfone.

Physical properties	
Density (g/cm^3)	1.24
Water absorption, 24 h (%)	0.30

Mechanical properties	
Tensile strength (psi)	10,200
Tensile modulus (psi)	360,000
Tensile elongation at break (%)	50–100
Flexural strength (psi)	15,400
Flexural modulus (psi)	390,000
IZOD impact notched (ft-lb/in)	1.3
Hardness	M75, R125, Shore D80

Thermal properties	
Coefficient of linear thermal expansion ($\times 10^{-5}$ in./in./°F)	3.1
Heat deflection temperature at 264 psi (°F)	345
Melting point (°F)	374
Maximum operating temperature (°F)	300

Electrical properties	
Dielectric strength, V/mil	425

Polysulfone (PSU) is a high temperature thermoplastic with high mechanical strength and rigidity. Some of the other properties include:
- Good creep strength over a wide range of temperatures
- Excellent dimensional stability
- Very good resistance to hydrolysis
- Good chemical compatibility
- Good mechanical strength
- Resistance to sterilization and stress cracking

7.5.2 Applications

PSUs are chemical inert and biocompatible and can be sterilized. These properties make them highly suitable for medical and food/beverage contact applications [14,15]. Many grades can withstand long-term exposure to hot chlorinated water and have received approval for food contact and drinking water.

Other applications of PSU include:
- Hemodialysis membrane
- Components of biopharmaceutical operation
- Housings
- Connector valves
- Clamps and sensor housing

7.5.3 Medical grades and suppliers

Some of the major suppliers in the PPSU market are listed below in Table 7.8.

7.6 Polyphenylene sulfide

7.6.1 Properties

Polyphenylene sulfide (PPS) is produced by reaction of sodium sulfide and dichlorobenzene in a polar solvent such as N-methylpyrrolidone and at a temperature of 250°C as shown in Fig. 7.7 [16].

Table 7.8 Major supplier in the polyphenyl sulfone market.

Supplier	Medical grade
Solvay	Udel PSU, Eviva PSU
BASF	Ultrason S
RTP Company	RTP 900 Series
UJU New Materials Co. Ltd.	Paryls PSU

Figure 7.7 Structure of polyphenylene sulfide (PPS) [16].

PPS is a semicrystalline, high-temperature thermoplastic polymer. Other properties of PPS include:
- Good resistance to chemicals
- Excellent electrical resistance
- Exceptional mechanical strength, even at temperatures above 200°C
- Very good dimensional stability
- Low water absorption
- Low susceptibility to creep

Table 7.9 shows physical, mechanical, thermal, and electrical properties of PPS.

7.6.2 Applications

PPS compounds are used in medical applications such as surgical instruments and device components and parts that require high dimensional stability, strength, and heat resistance. Fig. 7.8 shows surgical forceps handle part molded from Celanese Fortron polyphenylene sulfide (PPS) [17].

Table 7.9 Physical, mechanical, thermal, and electrical properties of PPS.

Physical properties	
Density (g/cm^3)	1.35
Water absorption, 24 h (%)	0.02
Mechanical properties	
Tensile strength (psi)	12,500
Tensile modulus (psi)	480,000
Tensile elongation at break (%)	4
Flexural strength (psi)	21,000
Flexural modulus (psi)	600,000
IZOD impact notched (ft-lb/in)	0.5
Hardness	M95, R125, Shore D85
Thermal properties	
Coefficient of linear thermal expansion ($\times 10^{-5}$ in./in./°F)	4.0
Heat deflection temperature at 264 psi (°F)	200
Melting point (°F)	536
Maximum operating temperature (°F)	424
Electrical properties	
Dielectric strength, V/mil	450

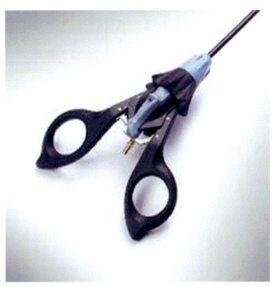

Figure 7.8 Surgical forceps handle part from Fortron polyphenylene sulfide (PPS) [17]. *Source: Image reproduced with permission from Celanese.*

Table 7.10 Major supplier in the polyphenyl sulfide (PPS) market.

Supplier	Medical grade
Solvay	Ryton PPS (R-4-242-NA, R-4-242-BL)
Celanese	Fortron MT PPS
SABIC	LNP LUBRICOMP Compound, LNP THERMOCOMP Compound

7.6.3 Medical grades and suppliers

Some of the major suppliers in the PPS market are listed in Table 7.10.

7.7 Polyvinylidene fluoride

7.7.1 Properties

Polyvinylidene Fluoride (PVDF) is synthesized by the free radical polymerization of 1,1-difluoroethylene ($CH_2=CF_2$) as shown in Fig. 7.9 [18]. The polymerization takes place in the suspension or emulsion from 10 to 150°C and pressure of 10–300 atm.

PVDF is an opaque, semicrystalline thermoplastic fluoropolymer. Some of the other properties include the following:
- Excellent chemical resistance
- Excellent mechanical strength

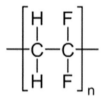

Figure 7.9 Structure of polyvinylidene fluoride (PVDF) [18].

- High dielectric strength
- Abrasion resistance
- Creep resistance
- Low moisture absorption
- Compatible with gamma sterilization

Table 7.11 shows physical, mechanical, thermal, and electrical properties of PVDF.

Table 7.11 Physical, mechanical, thermal, and electrical properties of polyvinylidene fluoride.

Physical properties	
Density (g/cm^3)	1.78
Water absorption, 24 h (%)	0.02
Mechanical properties	
Tensile strength (psi)	7800
Tensile modulus (psi)	350,000
Tensile elongation at break (%)	35
Flexural strength (psi)	10,750
Flexural modulus (psi)	310,000
IZOD impact notched (ft-lb/in)	3.0
Hardness	M75, R84, Shore D77
Thermal properties	
Coefficient of linear thermal expansion ($\times 10^{-5}$ in./in./°F)	7.1
Heat deflection temperature at 264 psi (°F)	235
Melting point (°F)	536
Maximum operating temperature (°F)	302
Electrical properties	
Dielectric strength, V/mil	350

7.7.2 Applications

Some of the application of PVDF includes [19,20]:
- Because of its superior wearability, chemical resistance, and inert characteristics, PVDF is used for coating and legibly marking identification in disposable medical devices and instruments.
- It is used as an ultra-fine fiber filtration system for high-efficiency face masks. Example includes Arkema's KYNAR®
- It is used in applications such as smart watch bands, protective cases, and internal components
- PVDF is used in minimally invasive medical catheters. Example includes Arkema's KYNAR® RX copolymer

7.7.3 Medical grades and suppliers

Although there are many players in the global PVDF market, there are only two major suppliers that are approved and operating in PVDF medical market as shown in Table 7.12:

7.8 Polyetherimide

7.8.1 Properties

Polyetherimide (PEI) as shown in Fig. 7.10 is produced by polycondensation reaction between bisphenol-A dianhydride such as tetracarboxylic dianhydride and a diamine such as m-phenylene diamine [21].

PEI is an amorphous polymer, and its properties include high mechanical strength and rigidity at elevated temperatures, long-term heat resistance,

Table 7.12 Major supplier in the polyvinylidene fluoride medical market.

Supplier	Medical grade
Arkema	KYNAR
Solvay	Solef PVDF

Figure 7.10 Structure of polyetherimide (PEI) [21].

Table 7.13 Physical, mechanical, thermal, and electrical properties of polyetherimide.

Physical properties

Density (g/cm^3)	1.27
Water absorption, 24 h (%)	0.25

Mechanical properties

Tensile strength (psi)	15,200
Tensile modulus (psi)	430,000
Tensile elongation at break (%)	60
Flexural strength (psi)	22,000
Flexural modulus (psi)	480,000
IZOD impact notched (ft-lb/in)	1.0
Hardness	M114, R123, Shore D86

Thermal properties

Coefficient of linear thermal expansion ($\times 10^{-5}$ in./in./°F)	3.1
Heat deflection temperature at 264 psi (°F)	392
Melting point (°F)	536
Maximum operating temperature (°F)	426

Electrical properties

Dielectric strength, V/mil	830

dimensional stability, and good electrical properties. Other key properties include the following:
- High creep resistance over a wide range of temperatures
- High permanent operating temperature
- Inherent flame retardant
- Resist hydrolysis
- Compatible with steam, EtO, and gamma radiation sterilization

Table 7.13 shows physical, mechanical, thermal, and electrical properties of PEI.

7.8.2 Applications

PEIs are the material of choice when long-term heat and chemical resistance and compatibility with sterilization are needed. Some examples include [22,23]:
- Stopcocks and pipettes
- Instrument and sterilization trays

- Containers, and accessories
- Medical monitor probe housing
- Surgical instrument handles and enclosures
- Nonimplant prosthesis

7.8.3 Medical grades and suppliers

There is only one major resin supplier in PEI medical market as shown in Table 7.14; however, there are many processors such as Rochling, Westlake Plastics (TEMPALUX) and others who uses Ultem to produce finished products.

Table 7.14 Major supplier in the polyetherimide medical market.

Supplier	Medical grade
SABIC	Ultem HU1000, Ultem HU1000E, Ultem HU1004, Ultem HU1010, Ultem HU1100, Ultem HU2100, Ultem 2110, Ultem 2200, Ultem 2210, Ultem 2300, Ultem 2310

7.9 Polydimethylsiloxane

7.9.1 Properties

Polydimethylsiloxane (PDMS) is produced by hydrolyzing chloromethyl silanes [(CH$_3$)$_2$SiCl$_2$], which is produced from high-purity SiO$_2$ and chloromethane [CH$_2$Cl$_2$] by the direct process. The structure of PDMS is shown in Fig. 7.11 [24].

Some of the key properties include:
- Biocompatibility
- Chemically inertness
- Low interfacial free energy

Figure 7.11 Structure of polydimethylsiloxane (PDMS) [24].

- Good gas permeability
- Good thermal stability
- Optically transparent

7.9.2 Applications

PDMS is a silicone elastomer. Some of the applications where PDMS is used include [25–27]:
- Catheter and drainage tubing
- Dialysis membrane
- Micropumps
- Microvalves
- Adaptive lenses

7.9.3 Medical grades and suppliers

The major suppliers for PDMS are listed in Table 7.15.

Table 7.15 Major suppliers in the polydimethylsiloxane medical market.

Supplier	Medical grade
Dow Corning	C6-7XX Series, SYLGARD 184
NuSil	MED-360, MED-361, MED-366
Momentive	TUFEL III 92506, LIM 6000 Series

7.10 Thermoplastic polyurethane

7.10.1 Properties

Thermoplastic Polyurethane (TPU) is produced via polyaddition reaction between a diisocyanate and polyols. As shown in Fig. 7.12, the structure of TPU is a linear segmented block copolymer composed of hard and soft segments. Soft segments are built from a polyol and an isocyanate, which provides flexibility and elastomeric character of a TPU. The hard segment is built from a chain extender and isocyanate giving TPU its toughness and physical performance properties [28].

Figure 7.12 Structure of thermoplastic polyurethane [28].

Table 7.16 Physical, mechanical, thermal, and electrical properties of thermoplastic polyurethane.

Physical properties	
Density (g/cm^3)	1.45
Water absorption, 24 h (%)	0.238

Mechanical properties	
Tensile strength (psi)	11,500
Tensile modulus (psi)	797,000
Tensile elongation at break (%)	86
Flexural strength (psi)	13,800
Flexural modulus (psi)	652,000
IZOD impact notched (ft-lb/in)	12.0
Hardness	R70, Shore D83

Thermal properties	
Coefficient of linear thermal expansion ($\times$ 10^{-5} in./in./°F)	31.0
Heat deflection temperature at 264 psi (°F)	44
Melting point (°F)	140
Maximum operating temperature (°F)	426

Electrical properties	
Dielectric strength, V/mil	400

TPU offers high durability and flexibility. Other key properties include:
- Abrasion and scratch resistance
- UV resistance
- High transparency
- Excellent mechanical properties
- Resistance against oils and chemicals

Table 7.16 shows physical, mechanical, thermal, and electrical properties of TPU.

7.10.2 Applications

TPU has been a material of choice in the medical industry because of its excellent mechanical properties, durability, and resistance against oils and chemicals.

Medical TPU applications include [29–31]:
- Diagnostic, anesthesia, and artificial respiration devices
- Healthcare mattresses

Table 7.17 Major suppliers in the thermoplastic polyurethane medical market.

Supplier	Medical grade
Covestro	DESMOPAN 9370
	TEXIN
BASF	ELASTOLLAN

- Dental materials
- Medical instrument cables
- Gel shoe orthotics
- Wound dressings

7.10.3 Medical grades and suppliers

The major suppliers for TPU are listed in Table 7.17.

7.11 Thermoplastic elastomer

7.11.1 Properties

Thermoplastic Elastomers (TPEs) are block copolymers which exhibit simultaneous thermoplastic and elastomeric properties. The polymer system has crystalline and amorphous domains. These two domains account for the unique properties of the TPE. The crystalline domain or the hard block is an orderly locked structure that gives the material thermoplastic properties. The amorphous domain, or the soft block, is a disordered structure that gives the elastomeric properties.

TPE is safe and biocompatible and has irritation-free properties. Other key properties include:
- Excellent flexural fatigue resistance
- Good tear and abrasion resistance
- High impact strength
- Excellent resistance to chemical and weathering
- Possesses low compression set
- Excellent barrier properties
- Compatible with different sterilization methods i.e., steam, EtO, and gamma radiation

Table 7.18 shows physical, mechanical, thermal, and electrical properties of TPE.

Table 7.18 Physical, mechanical, and thermal properties of thermoplastic elastomer.

Physical properties	
Density (g/cm^3)	0.95
Mechanical properties	
Tensile strength (psi)	1740–4210
Tensile elongation at break (%)	330–560
Thermal properties	
Heat deflection temperature at 264 psi (°F)	105–275
Melting point (°F)	340–430

7.11.2 Applications

TPEs are used in applications that require flexibility, strength, chemical resistance, and inertness to chemicals. TPE is frequently used as a material in oxygenation masks [32].

Some of the other examples include [33]:
- Medical tubes
- Syringe baskets
- Uterine balloon tamponades
- Oxygen masks
- Surgical elastomer wipes
- Disposable surgical gowns
- Endoscopy biopsy valve

7.11.3 Medical grades and suppliers

The major suppliers for TPE are listed in Table 7.19.

Table 7.19 Major suppliers in the thermoplastic elastomer medical market.

Supplier	Medical grades
Teknor Apex	Medalist
Kraiburg	THERMOLAST
RTP Company	Polabond
DSM	Arnitel
Avient	Versaflex HC

References

[1] S.A. Ashter, Polyacetals, pg. 53, chapter 3: mechanisms of polymer degradation, in: Introduction to Bioplastics Engineering, Elsevier, 2016.
[2] Polyacetals, Polymer Properties Database. https://polymerdatabase.com/polymer%20classes/Polyacetal%20type.html.
[3] Polyacetal: Detailed Information on POM and its Features, Omnexus. https://omnexus.specialchem.com/selection-guide/polyacetal-polyoxymethylene-pom-plastic.
[4] Acetal Copolymer, Poly-Tech Industrials. www.polytechindustrial.com/products/plastic-stock-shapes/acetal-copolymer.
[5] Acetal, Dielectric Manufacturing. https://dielectricmfg.com/knowledge-base/acetal/.
[6] D. Parker, J. Bussink, H.T. Van de Grampe, G.W. Wheatley, E.-U. Dorf, E. Ostlinning, K. Reinking, Polymers, high-temperature, in: Ullmann's Encyclopedia of Industrial Chemistry, 2012.
[7] D. Kemmish, Update on the Technology and Applications of Polyaryletherketones, iSmithers Rapra Publishing, 2010.
[8] Polyetheretherketone, Milton Plastics. http://www.miltonplastics.com/index.php/Picture/show/10.html.
[9] What is polyetheretherketone, Fluorocarbon. https://fluorocarbon.co.uk/news-and-events/post/13/what-is-polyethererketone-peek.
[10] Polyetheretherketone, Dielectric Manufacturing. https://dielectricmfg.com/knowledge-base/peek/.
[11] N. Sparrow, Twin PEEKs: World-First for Invibio; Evonik Debuts Radiolucent Implant-Grade Material, Plastics Today, 2019. www.plasticstoday.com/medical/twin-peeks-world-first-invibio-evonik-debuts-radiolucent-implant-grade-material.
[12] Polyphenylsulfone, Omnexus. https://omnexus.specialchem.com/selection-guide/polyphenylsulfone#:~:text=Polyphenylsulfone%20(PPSU)%20is%20a%20transparent,temperature%20of%20274%C2%B0C.
[13] Polysulfone, Polymer Properties Database. https://polymerdatabase.com/polymer%20classes/Polysulfone%20type.html.
[14] Udel® PSU, Solvay. www.solvay.com/en/brands/udel-psu/applications.
[15] L.M. Sherman, PSU Outperforms PC in Sanitary Biotech Sight Gauge, Plastics Technology, 2021. https://www.ptonline.com/articles/psu-outperforms-pc-in-sanitary-biotech-sight-gauge->.
[16] Polyphenylene Sulfide, Omnexus. https://omnexus.specialchem.com/selection-guide/polyphenylene-sulfide-pps-plastic-guide.
[17] Fortron® PPS, High-Temperature Stability, Broad Chemical Strength and Creep Resistance, Celanese. www.celanese.com/en/engineered-materials/products/fortron-pps.
[18] Polyvinylidene Fluoride, Omnexus. https://omnexus.specialchem.com/selection-guide/polyvinylidene-fluoride-pvdf-plastic.
[19] KYNAR® PVDF for High Efficiency Face Masks (N95, FFP1, FFP2, FFP3, KN95), Arkema. www.extrememe materials-arkema.com/en/markets-and-applications/consumer-goods-and-healthcare/kynar-pvdf-for-face-masks/.
[20] Arkema Introduces Kynar RX Copolymer for Medical, Med Device Online, 2013. www.meddeviceonline.com/doc/arkema-introduces-kynar-rx-copolymer-for-medical-0001.
[21] Polyetherimide, Omnexus. https://omnexus.specialchem.com/selection-guide/polyetherimide-pei-high-heat-plastic.
[22] Boekel® Ultem® 1000 Round Plastic Pipette Storage and Sterilization Cannister. www.capitolscientific.com/Boekel-135906-Ultem-1000-Round-Plastic-Pipet-Storage-Sterilization-Cannister-15-2cm-Tall.
[23] Probe Housing, RTP Company. www.rtpcompany.com/probe-housing/.

[24] Introduction to Polydimethylsiloxane, Elveflow. www.elveflow.com/microfluidic-reviews/general-microfluidics/the-polydimethylsiloxane-pdms-and-microfluidics/.
[25] Silicone PDMS Membrane, Interstate Specialty Products. www.interstatesp.com/products/membrane-filters/silicone-membranes/.
[26] High Purity Silicones, NuSil®, Avantor Sciences. www.avantorsciences.com/pages/en/nusil-high-purity-silicones.
[27] Catheters, Drainage, and Suction Devices, Momentive. www.momentive.com/en-us/industries/health-care/catheters.
[28] Thermoplastic Polyurethanes, Omnexus. https://omnexus.specialchem.com/selection-guide/thermoplastic-polyurethanes-tpu.
[29] D. Sanchez, Innovative Soft Thermoplastic Urethanes for Medical Applications, Medical Design Briefs, 2012. www.medicaldesignbriefs.com/component/content/article/mdb/features/articles/12814.
[30] Wound Care: High Tech Materials for Medical Applications, Covestro. www.solutions.covestro.com/en/highlights/articles/theme/applications/wound-care.
[31] Elastollan® for Medical Applications, Performance Polymers, BASF. https://plastics-rubber.basf.com/global/en/performance_polymers/products/Medical-Applications-Elastollan.html.
[32] A. Esposito, First Nasal-Only Oxygenation Mask Uses Super-soft Medical TPE for Strong, Cushioning Seal to Patient Face, Medical Design Nd Outsourcing, 2016. www.medicaldesignandoutsourcing.com/first-nasal-only-oxygenation-mask-uses-super-soft-medical-tpe-for-strong-cushioning-seal-to-patient-face/.
[33] W.S. Ripple, J. Simons, Thermoplastic Elastomers in Medical Devices, Technical Contribution to MedPlast Supplement, GLS Corporation, IL, 2007.

CHAPTER EIGHT

Plastics fabrication techniques

Contents

8.1 Machining of plastics	129
8.1.1 Introduction	129
8.1.2 Methods of machining plastics materials	130
8.1.2.1 Computer numerical control machining	130
8.1.2.2 Plasma Cutters	134
8.1.2.3 Electric discharge machining	134
8.1.2.4 Turning	136
8.1.2.5 Water jet cutting	137
8.1.2.6 Other processes	138
8.2 Bonding	139
8.2.1 Mechanical fastening	139
8.2.2 Solvent bonding	139
8.2.3 UV bonding	140
8.2.4 Ultrasonic welding	140
8.2.5 Adhesive bonding	140
8.3 Staking	141
8.3.1 Introduction	141
8.3.2 Types of staking	142
8.3.2.1 Thermal staking	142
8.3.2.2 Flared staking	142
8.3.2.3 Spherical staking	142
8.3.2.4 Flush staking	143
8.3.2.5 Hollow staking	143
8.3.2.6 Knurled staking	143
8.4 Two-part molding (silicone molding)	143
References	144

8.1 Machining of plastics
8.1.1 Introduction

Plastics are versatile materials that have found usage in a range of products. One of the main characteristics of plastics is their ability to be molded into a finished product with no secondary work to be carried out. Complicated shapes, holes, and undercut features can be molded into the

Applications of Polymers and Plastics in Medical Devices
ISBN: 978-0-12-820980-6
https://doi.org/10.1016/B978-0-12-820980-6.00013-8
© 2022 Elsevier Inc.
All rights reserved.

component using molding processes such as injection molding, blow molding, compression molding, transfer molding, and so on. However, parts suffer from problems such as warpage, sink marks, weld lines, and poor surface finish that affect part tolerance and dimensional accuracy [1,2].

Injection molding process is typically used in high volume production of parts. It is not preferable to produce smaller quantities due to high cost of making injection mold and the waste that is generated from runners. Therefore, production of small quantities of products is primarily done by machining process [3]. Not all plastic materials can be machined. The more rigid a plastic then the easier it is to be machined. The more flexible and the softer plastics are not suitable for machining [4].

The cutting tools used in the machining of all materials rely on the rigidity of the component being cut. In the case of cutting metals, the materials' natural rigidity is good. Therefore, the component resists distortion when the cutter (saw, drill, or machine bit) cuts the metal. In the case of plastics, machining tends to lend itself better to rigid materials, such as fiber reinforced thermosetting plastics materials, glass reinforced nylons, acrylic, or PEEK, which have good relative stiffness. Less rigid plastic tends to deform and bend away when the cutter attempts to cut the component, making the achievement of fine dimensional tolerances difficult [4]. Table 8.1 shows advantages and disadvantages of machining plastics [4].

Some important considerations are required when machining plastics components [4]:
1. Machining fixtures should be designed with jaws that can be used as a grip to hold plastic components while they are being machined.
2. If the component being machined is thermoplastic, it can be cooled by air if the resultant swarf is continuous and not in chipping form.
3. When a component is machined, the heat is generated from the machining process, which causes component to undergo thermal expansion upon cooling. Therefore, it is important to take thermal expansion into consideration when machining.

8.1.2 Methods of machining plastics materials
8.1.2.1 Computer numerical control machining
Computer numerical control or referred more commonly by its acronym CNC is an automated control of machining tools by a computer [5]. When a part is to be machined, its profile is programmed into a computer. This software then dictates corresponding tools to perform tasks in the order they were programmed. If the component to be cut has a complex shape, its

Plastics fabrication techniques

Table 8.1 Advantages and disadvantages of machining plastics.

Advantages	Disadvantages
• No mold costs are needed	• Machining ability limited to the more rigid plastics materials
• Ability to manufacture plastic components with short lead times	• Relative high cost of block plastic material
• Ability to manufacture low volumes economically	• High scrap (relative to other plastics forming processes) can result
• Can trial a design before committing to tooling	• High volume of swarf to be removed can present difficulties
• Thicker wall sections can be accommodated	• High costs of CNC machine time
• Components too large to be molded can be machined from fabricated plastic	• Volume production by machining will require robust jigs and fixtures
• The forces required to machine plastics are low	• Plastics materials do not conduct away any heat generated in the machining process
• Plastics normally machine dry	• Dust-producing composite plastics require an effective dust collection system
• Swarf can be recycled back into the compounding process	

profile can be programmed into a computer. A CNC machining center can be used to manufacture duplicate numbers of components. Multiple interchangeable cutters typically used on CNC machines enable complex and varied components to be machined [6].

Position control is determined through an open-loop or closed-loop system. With the former, the signaling runs in a single direction between the controller and motor. With a closed-loop system, the controller can receive feedback, which makes error correction possible. Thus, a closed-loop system can rectify irregularities in velocity and position.

In CNC machining, movement is usually directed across X and Y axes. The tool, in turn, is positioned and guided via stepper or servo motors, which replicate exact movements as determined by the G-code. If the force and speed are minimal, the process can be run via open-loop control. For everything else, closed-loop control is necessary to ensure the speed, consistency, and accuracy required for industrial applications, such as metalwork [6].

Computer numerical control mills

CNC mills are computer-controlled milling machines that progressively remove material from the workpiece using the cutting to produce the desired part. These machines can perform variety of operations such as drilling, turning, and other mechanical processes on a wide range of materials including plastic, wood, glass, and metal. They come in two machine configurations—(1) horizontal and (2) vertical and are differentiated based on the number of axis system (X, Y, or Z).

Each mill has a machine interface, column, knee, saddle, worktable, spindle, arbor, ram, and a machine tool. A wide range of machine tools can be used depending on the milling application. Depending on the machine configuration, the machine spindle can be either vertically or horizontally oriented. Depending on the required application, CNC mills can have between 2- and 5-axis of motion. A 2-axis machine provides horizontal movement in X- and Y-axis. X-axis corresponds to side-to-side, while Y-axis is forward-and-back movement on a flat plane. In mills with three or more axis, X- and Y-axis designate horizontal movement, while Z-axis represents vertical movement, and the W-axis represents diagonal movement across a vertical plane [7].

There are four primary types of CNC milling machines [7]:
(a) Vertical mill
(b) Turret mill
(c) Bed mill
(d) Horizontal mill

Table below shows comparison between different types of CNC milling machines. Each machine differs from one another based on the orientation of the spindle and how the cutter is positioned on the spindle. For example, in a Vertical mill as shown in Figure 8.1 [8], the spindle is vertically oriented while in the Turret mill, spindle remains stationary and the table underneath the spindle moves [7].

Figure 8.1 An example of vertical CNC mill [8].

Vertical mill	Turret mill	Bed mill	Horizontal mill
• Spindle is vertically oriented • Cutters positioned on the spindle	• Spindle is stationary while the table below it moves perpendicular and parallel to the spindle axis • Quill allows cutter to move up and down • Allows vertical cuts • Most effective when the machine is small	• Spindle is stationary while the table below it moves perpendicular to the spindle axis • Considered more rigid machines than turret mills	• Spindle is vertically oriented • Cutters are mounted on a horizontal arbor

Computer numerical control lathes

A CNC lathe is a machine tool where the material or part is held in place and rotated by the main spindle as the cutting tool that works on the material mounted and moved in various axes [7,9]. These machines perform variety of operations such as facing, tapering, and turning. CNC lathes come in five machine configurations—(1) horizontal, (2) vertical, (3) slant bed, (4) flat bed, and (5) standard and are differentiated based on the number of the axis system that can range from 2- to 5-axis [10].

Depending on the configuration, lathes can consist of machine interface, machine bed, spindle system, chuck, guideway, headstock, tailstock, slides, and turret [10]. A turret is the section of the machine that holds the tool holders and indexes them. A work piece is held by the spindle. There are slides that allow the turret to move in multiple axes at the same time [9]. CNC lathes can be programmed either directly on the machine using CNC programming system or offline using CAD/CAM system. In a typical CNC lathe operation, only one program can be processed while the next program is being established to save time and increase productivity.

There are four primary types of CNC lathes [11]:

(a) Speed lathe
(b) Engine lathe
(c) Turret lathe
(d) Tool room lathe

Comparison between different types of CNC milling lathes are shown in the Table below [12]:

Speed lathe	Engine lathe	Turret lathe	Tool room lathe
• Simple design • Only consists of headstock, tailstock, and tool post • Operate in four speeds • Used for light machine work such as metal spinning, wood turning and metal polishing	• Most popular lathe • Excellent for lower power operation • Comes in variety of lengths	• Used for quick operations • Capstan wheel for positioning the next tool • Performs a sequential process using turret without moving the workpiece	• Very versatile lathe • Provides excellent finish • Capable of running at higher speeds and providing different feeds

8.1.2.2 Plasma Cutters

Plasma cutter is a machine that cuts metal by sending air or an inert gas through a plasma torch. This generates an electrical arc that can reach temperatures up to 45,000°F forcing the plasma through the torch tip and cutting the metal [13]. The process is foremost applied to metal materials but can also be employed on other surfaces. To have the speed and heat necessary to cut metal, plasma is generated through a combination of compressed-air gas and electrical arcs [6]. Fig. 8.2 shows Robotworx plasma cutter [14].

8.1.2.3 Electric discharge machining

Electric discharge machining (EDM) also known by other names such as spark machining die sinking and arc machining is a process of removing portions of a workpiece using electric discharge. In this process, an electrical current is passed between electrodes and a workpiece, which is separated by a dielectric fluid. At the start of the process, the dielectric fluid acts as an insulator, however, when voltage reaches its ionization point, it acts as an electrical conductor causing spark discharge resulting in portions of work piece removed to form a desired shape [6,15].

There are three primary types of EDM machines (Fig. 8.3) [16,17]:
1. Die sinker EDM
2. Wire EDM
3. Hole drilling EDM

Plastics fabrication techniques

Figure 8.2 Robotworx plasma cutter [14].

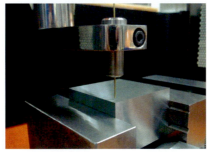

Figure 8.3 Examples of electric discharge machines (EDM) machining (EDM) machines [16,17].

Comparison between different types of EDM machines are shown in Table below [15]:

Die sinker EDM	Wire EDM	Hole drilling EDM
• Used for creating complex shapes • An electrode and work piece are soaked in dielectric fluid for the purpose of piece formation • Process starts with machining a graphite electrode forming a positive electrode. Electrode is then plunged into the work piece causing sparking over its surface • Uses hydrocarbon oil for the dielectric fluid	• Spark erosion is used to remove portions from an electronically conductive material • Uses a thin wire for an electrode • Wire moves in a controlled pattern generating spark between the wire and the workpiece • Uses a spool of wire to present fresh discharge path • Uses deionized water for the dielectric fluid	• Produces excellent surface finish and minimal heat affected zones • Ability to cut hardened materials and exotic alloys • Uses rotating conductive tube for electrode and deionized water for dielectric fluid to flush the cut • Used for creating pilot holes for wire threading

Dielectric fluid serves three essential functions:
- Controlling the spacing of the sparking gap between the electrode and workpiece
- Cooling the heated material to form the EDM chips
- Removing EDM chips from the sparking area

EDM process produces chips that are composed of material from both the electrode and the workpiece. These materials need to be removed from the cutting zone, which is accomplished by flushing the dielectric fluid through the sparking gap [6,15]. Over a period, the dielectric fluid breaks down either by contamination or as a result of age. This increases the risk of unstable discharge. This requires monitoring and continuous circulation of clean dielectric fluid through the cutting zone to flush impurities.

8.1.2.4 Turning

Turning is a machining process used to make cylindrical parts, where the cutting tool moves in a linear fashion while the work piece rotates. A turret that is attached to the tool is programmed to move to the workpiece and remove material to create the programmed result. During the CNC turning process, the materials such as metal, plastic, and wood are rotated with a CNC. CNC turning can be performed on the outside of the workpiece or the inside (also known as boring) to produce tubular components to various geometries [18].

Some additional considerations that are required when using turning operation [18,19]:
- Fine, C-2 grade carbide inserts are recommended for turning.
- Polished top surfaces will help to reduce material build-up, allowing for better surface finishes.
- Cutting edges should have generous relief angles and negative back rake to minimize any rubbing action.

There are four different types of turning:
1. Step turning
2. Taper turning
3. Chamfer turning
4. Contour turning

Comparison between different types of turning machines are shown in Table below [18,19]:

Step turning	Taper turning	Chamfer turning	Contour turning
• Produces step-like feature by creating two surfaces	• Produces a ramp like transition between the two surfaces with different diameters • Diameter of the workpiece changes uniformly from one end to another	• Creates angled transition between the two surfaces with different tuned diameters	• Cutting tool axially follows the path with a predefined geometry • Multiple passes of a contouring tool are necessary to create desired contours

8.1.2.5 Water jet cutting

Water jet cutting is a process of cutting a wide array of materials using an extremely high-pressure jet of water or a mixture of water and an abrasive substance. For example, it is used to edge trim fiber-reinforced thermosetting components, which would otherwise prove difficult to trim by other processes. A primary advantage of waterjet cutting over some other operations is that it adds no heat to the material during the process, minimizing the need for postcutting heat treatment [20].

Water jet cutting machine consists of following components [20]:
- CNC guiding machine
- High-pressure pump and piping

An example of water jet cutter is shown in Figure 8.4 [21]:

Figure 8.4 Water jet cutter [21].

- Pressure intensifier
- Electric motor
- Nozzle with varying diameter of sizes
- Filtration system

Water jets are employed as a cooler alternative for materials that are unable to bear the heat-intensive processes of other CNC machines. As such, water jets are used in a range of sectors, such as the aerospace and mining industries, where the process is powerful for the purposes of carving and cutting, among other functions. Water jet cutters are also used for applications that require very intricate cuts in material, as the lack of heat prevents any change in the materials' intrinsic properties that may result from metal on metal cutting [21].

8.1.2.6 Other processes
Sawing
In this method of machining, sections of plastic material are parted off from the workpiece for other machining operations. Thick-walled parts are sawed with relatively thin blades to avoid excessive frictional heat generation. It is recommended to use well-sharpened, strongly offset blades. For cutting engineering plastic materials such as nylon and acrylic, it is recommended to use specially designed blades [22].

Die cutting

Die cutting is limited to sheet material and produces a simple component. In this process, a male and female dies are used to punch out a predetermined shape. This process can be either a manual process or automated using a special machine.

Hot knife cutting

This method of cutting is reserved for less rigid plastics that can be cut using a hot knife to slice. An electrically heated wire or blade melts the plastic locally. This type of process is commonly used to cut blocks of foam and expanded polystyrene.

Laser cutting

This process can be used for cutting and profile boring of certain types of acrylic and other plastics although not thermosetting. This process uses light to direct energy on the material to be processed. Complex angled cuts are feasible using laser cutting process.

8.2 Bonding

Plastic bonding provides an alternate solution to adhesive for plastic manufacturers who are looking for efficient and permanent way to join plastic components into complete assemblies [23]. There are five main methods to bond plastics [24]:
1. Mechanical fastening
2. Solvent bonding
3. UV bonding
4. Ultrasonic welding
5. Adhesive bonding

8.2.1 Mechanical fastening

Typical mechanical fasteners include rivets, screws, nuts, or pins. These are components that can be put in place, glued, forced, or expanded into holes. It is important that both the plastic component and the fastener being used are strong to withstand forces during installation [24].

8.2.2 Solvent bonding

Solvent bonding is a two-step process in which the two plastic components are first softened with a coating of solvent and then they are clamped together under applied pressure. Both components are held together for 24–48 h under room temperature as the solvent dries and evaporates. It is important to note that excess applied pressure may cause the components to distort [24].

8.2.3 UV bonding

UV bonding is an efficient way to bond plastic components using a high-intensity ultraviolet light. UV bonding works in seconds to bond plastic components to a wide range of materials such as ceramic, glass, and metal. This process provides many advantages such as reduced rejection rates, increased production speed, improved solvent, and scratch resistance, as well as superior bonding [24].

8.2.4 Ultrasonic welding

Ultrasonic welding utilizes sonic pulses that are directed to the weld area with a resonant vibrating device called a horn. The horn causes the two plastic components to vibrate against one another, and the vibration results in heat that fuses the elements together. Thus, no solvents or glue are required.

Plastic components and other products, including alloys and blends of a variety of resin families, can be fastened by ultrasonic welding. Moreover, different materials may be joined if the melting temperatures are no more than 30° apart and the compositions are compatible.

8.2.5 Adhesive bonding

Whether bonding plastic to plastic or other materials, adhesives offer several benefits over other joining methods. Adhesives distribute loads evenly over a broad area, reducing stress on the joint. Because adhesives are applied inside the joint, they are invisible within the assembly.

Adhesives resist flex and vibration stresses and form a seal as well as a bond, which can protect the joint from corrosion. They join irregularly shaped surfaces more easily than mechanical or thermal fastening, barely increase assembly weight, create virtually no change in part dimensions or geometry, and quickly and easily bond dissimilar substrates and heat-sensitive materials.

Some of the limitations of adhesives include the following:
- Adhesives require setting and curing time for the adhesive to fixture and strengthen fully.

- Adhesives require preparation prior to assembly and may not work well when there is a need to repeatedly assemble and disassemble the joint.
- Adhesive should be compatible with the substrate and the operating environment.

8.3 Staking

8.3.1 Introduction

Plastic staking is a method of joining components together that uses a molded stud or boss to mechanically retain a mating component. Heat is applied to the boss, softening it. A forming tool is then used to reshape the material into a cap or stake [25].

In a staking process, heat and pressure can be used to join two or more components to one another. This creates a solid, hardware-free bond that uses the plastics' inherent strength to keep the finished assembly together. This process does not create mechanical stress because of the control and application of the temperature and pressure. The heating is completely localized to a small area [26].

Applications for heat staking include the following [27]:
- Polymer to polymer to other materials
- Flat head tip welding
- Sealing holes
- Embedding parts
- Polymer to mesh
- Breathable membrane
- Rim swaging
- Bonding
- Polymer to polymer welding
- Polymer to polymer

Staking process can be used with a range of materials including the following plastics:
- Thermoplastics
- Glass-filled thermoplastics
- Polycarbonate (PC)
- Polypropylene (PP)
- Polystyrene (PS)
- Acrylonitrile butadiene styrene (ABS)
- Nylon

8.3.2 Types of staking

Staking can be divided into six different types:
1. Thermal staking
2. Flared staking
3. Spherical staking
4. Flush staking
5. Hollow staking
6. Knurled staking

8.3.2.1 Thermal staking

Thermal staking process is often used to weld two parts consisting of dissimilar materials, which cannot be welded otherwise. It is also referred to as heading or riveting process where the controlled flow of plastic is used to capture or retain another component [27].

The integrity of a thermally staked assembly depends greatly upon the geometric relationship between the boss and the thermal tip. Proper design will produce optimum strength with minimum flash. Thermal staking should be considered when the parts to be assembled are still in the design stage. Several configurations for boss/cavity design are available, each with specific features and advantages. Their selection is determined by such factors as type of plastic, part geometry, assembly requirements, machining and molding capabilities, and cosmetic appearance [27].

Commonly used application involves the attachment of metal to plastic. A hole in the metal part receives a premolded plastic boss. The thermal tip contacts the boss and creates localized heat. As the boss melts, the light pressure from the horn forms a head to a shape determined by the horn tip configuration. When the heat is terminated, the plastic material solidifies, and the dissimilar materials are fastened together.

8.3.2.2 Flared staking

The low and high-profile flared staking satisfies the requirements of most applications. This staking is recommended for bosses with an outer diameter (OD) of 1.6 mm or larger and is ideally suited for low-density, nonabrasive amorphous plastics [27].

8.3.2.3 Spherical staking

The low and high-profile spherical staking is preferred for bosses with an outer diameter (OD) less than 1.6 mm and is recommended for rigid crystalline plastics with sharp, highly defined melting temperatures for plastics with abrasive fillers and for materials that degrade easily [27].

8.3.2.4 Flush staking
The flush staking is used for applications requiring a flush surface. The flush staking requires that the retained piece has sufficient thickness for a chamfer or counterbore [27].

8.3.2.5 Hollow staking
Bosses with an outer diameter (OD) more than 4 mm should be made hollow. Staking a hollow boss produces a large, strong head without having to melt a large amount of material. In addition, the hollow staking prevents sink marks on the opposite side of the component enabling parts to be reassembled with self-tapping screws in the event repair and/or disassembly is needed [27].

8.3.2.6 Knurled staking
The knurled staking is used in applications where appearance and strength are not critical. Since alignment is not an important consideration, the knurled staking is ideally suited for high-volume production. Knurled tips are available in a variety of fine, medium, and coarse configurations [27].

8.4 Two-part molding (silicone molding)

Liquid injection molding (LIM) also known as two-part molding is a process that initiates with a two-part liquid silicone material [28]. Part A and part B are delivered to a static mixer at 1:1 ratio from material drum and pail by a hydraulic and pneumatic metered pumping system. The static mixer blends the two components into a homogeneous suspension. This mixing process activates the curing system. The liquid silicone material from the static mixer then flows to the injection unit [29].

The liquid silicone rubber (LSR) material is then injected into the mold cavity through a runner and gate system where it is held in the mold under high pressure and elevated temperature until material is cured. Once the silicone material is inside the barrel, a shot of the cool mixed silicone advances to the mold as the nozzle seals firmly against the mold. The nozzle shut-off valve opens, and a measured shot of cool liquid silicone is injected into a hot (275—390°F) clamped mold. Then the nozzle shut-off valve opens, and the barrel retracts from the mold, while the screw begins to build another measured shot of cold liquid silicone. The cycle time is established to reach an optimal level of cure [29].

The following table shows advantages and disadvantages of LIM process [29]:

Advantages	Disadvantages
• Shorter cure times compared to traditional compression molding • High level of repeatability, good for tight tolerance/precision components • Superior clarity, material can be pigmented in-line with material flow to produce colors • Closed mold injection supports molding of complex geometries and overmolding	• Higher start-up/shutdown costs, better suited for high volume applications • Runner systems can lead to increased gross material weights when cold runner systems or other low waste options are not utilized

References

[1] J.A. Brydson, Plastics Materials, fifth ed., Butterworth Scientific, London, 1989.
[2] T.U. Jagtap, H.A. Mandave, Machining of plastics: a review, Int. J. Eng. Res. Gen. Sci. 3 (2 (2)) (2015).
[3] R.J. Crawford, Plastic Engineering, third ed., Butterworth Heinemann, 1998.
[4] British Plastics Federation, Machining of Plastics. https://www.bpf.co.uk/plastipedia/processes/Machining_of_Plastics.aspx.
[5] A. Overby, CNC Machining Handbook, McGraw Hill, 2011.
[6] B. Hess, What is CNC Machining? An Overview of the CNC Machining Process, Astro Machine Works, 2017. https://astromachineworks.com/what-is-cnc-machining/.
[7] M&M Automatic Products, Inc., Types of CNC Milling Machines, 2015. https://www.mmautomatic.com/types-of-cnc-milling-machines/.
[8] G. McKechnie, Wikipedia Commons Media Repository, 2005. https://commons.wikimedia.org/wiki/File:DeckelMaho-DMU50e-MachiningCenter.
[9] CNCCOM, What is a CNC Lathe and How Does it Work, 2015. https://www.cnc.com/what-is-a-cnc-lathe-and-how-does-it-work/.
[10] Hwacheon, Introduction to CNC Lathes. https://hwacheonasia.com/cnc-lathes/.
[11] T. Chuen Yap, V.C. Venkatesh, S. Izman, N.S. M El-Tayeb, P. Brevern, Cryogenic Facing of Ti-6AI-4V Alloys, International Conference on Manufacturing Science and Technology, Melaka, Malaysia, 2006.
[12] NexGen Machine Company, Understanding Different Types of CNC Lathe Machine, 2017. http://www.nexgenmachine.com/understanding-different-types-of-cnc-lathe-machine/.
[13] Bakers Gas, How to Choose a Plasma Cutter. https://bakersgas.com/pages/how-to-choose-a-plasma-cutter.
[14] Robotworx, Wikimedia Commons Repository, 2008. https://en.wikipedia.org/wiki/Plasma_cutting.
[15] I. Wright, EDM 101: Electrical Discharge Machining Basics, Engineering.com, 2017. https://www.engineering.com/story/edm-101-electrical-discharge-machining-basics.

[16] G. McKechnie, Wikimedia Commons Repository, 2005. https://commons.wikimedia.org/wiki/File:Robofil-300-WireCut.
[17] Wikimedia Commons Repository, 2011. https://commons.wikimedia.org/wiki/File:Small_hole_drilling_EDM_machines.
[18] Pioneer Services Inc., What is CNC Turning, 2016. https://pioneerserviceinc.com/blog/what-is-cnc-turning/.
[19] G. Immerman, CNC Turning and Turning Center Basics, Machine Metrics. https://www.machinemetrics.com/blog/cnc-turning-center-basics.
[20] S. Donath, Water Jet Cutting — Function, Method and Application Examples, Cutting Technology, 2019. https://www.etmm-online.com/water-jet-cutting–function-methods-and-application-examples-a-834966/.
[21] Wikipedia Commons Media Repository, 2019. https://em.wikipedia.org/wiki/Numerical_control#/media/File:Waterjet_cutting_machine.
[22] D. John, Beadle, Fabricating Plastics, Macmillan Education Ltd, 1972, pp. 18—19.
[23] P.J. Courtney, Guidelines for Bonding Plastics, Machine Design, 2000. https://www.machinedesign.com/fastening-joining/article/21834657/guidelines-for-bonding-plastics.
[24] CraftTech Industries, 4 Essential Techniques for Bonding Plastic Components. https://www.craftechind.com/4-essential-techniques-for-bonding-plastic-components/.
[25] Extol, Staking. https://www.extolinc.com/learning/staking/.
[26] Emerson, Technology for Specialized Applications. https://www.emerson.com/en-us/automation/welding-assembly-cleaning/heat-staking.
[27] Thermal Press International Inc. Why Use Heat Staking for Plastic Assembly Projects. https://www.thermalpress.com/why-use-heat-staking-for-plastic-assembly/.
[28] Silicone Technology Corp, Different Method of Molding Silicone Rubber, 2015. https://www.sitech-corp.com/blog/different-methods-of-molding-silicone-rubber/.
[29] Datwyler Sealing Solutions USA, Inc. http://usa.datwyler.com/liquid-injection-molding.html.

CHAPTER NINE

Plastic medical device manufacturing processes

Contents

9.1 Introduction	147
9.2 Extrusion	148
9.2.1 Background	148
9.2.2 Tubing extrusion	150
9.2.3 Film extrusion	150
9.3 Injection molding process	151
9.4 Catheter manufacturing process	153
9.4.1 Extrusion process	153
9.4.2 Secondary operations	156
9.4.2.1 Catheter shaft drawdown	*156*
9.4.2.2 RO marker swaging	*157*
9.4.2.3 Proximal and distal skiving	*157*
9.4.2.4 Balloon proximal and distal thermal welding	*157*
9.4.2.5 Proximal fuse	*157*
9.4.2.6 Manifold bonding	*158*
9.4.2.7 Balloon pleating and folding	*158*
9.5 Other secondary operations	158
9.5.1 Laser marking	158
9.5.2 Lubrication	159
9.5.3 Siliconization	159
9.5.3.1 Silicone	*159*
9.5.3.2 Metals	*160*
9.5.3.3 Glass	*161*
9.5.3.4 Plastics	*162*
References	162

9.1 Introduction

Medical device manufacturing includes all aspects of the fabrication of a medical device, from designing a manufacturing process to scale up to ongoing process improvements. It also includes the sterilization and packaging of a device for shipment. Extrusion and injection are two major manufacturing processes that have found use in plastic manufacturing. This chapter will discuss extrusion and injection molding processes with respect to medical device manufacturing. In addition, catheter manufacturing process will also be discussed.

9.2 Extrusion

9.2.1 Background

Extrusion is a polymer processing technique which is used either to form or to transport the molten plastic before it is formed. It is extensively used and can be understood by meat grinding example. Like most plastics are processed using screw extrusion, the meat grinder takes in large chunks of meat and uses in-built screw to reduce the size and mix it all up, and then extrude out lean strands of meat from the face of the meat grinder.

Although there are many types of extruders, the most common is the single-screw extruder as shown in Fig. 9.1 [1].

- This extruder consists of a screw in a metal cylinder or barrel. One end of the screw is connected through a thrust bearing and gear box to a drive motor that rotates the screw in the barrel. The other end is free floating in the barrel.
- One end of the barrel is connected to the feed throat. This feed throat is a separate barrel section, which is typically cooled with water, contains an opening called a feed port, and is connected to the feed hopper.

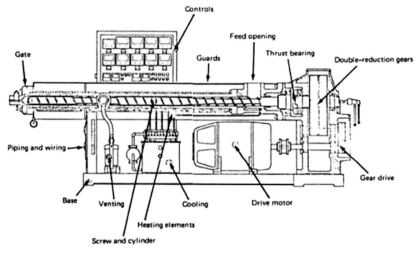

Figure 9.1 Typical single-screw extruder [1].

- The opposite end of the barrel is fully open and exposes the tip of the screw. A die adapter is usually connected to the open end of the extruder. A breaker plate and a screen pack are sandwiched between the extruder and die adaptor. The breaker plate provides a seal between the extruder and die and converts the rotational motion of the melt (in the extruder) to linear motion (for the die) and supports the screen pack. The screen pack filters the melt, thereby preventing unmelted resin, degraded polymer, and other contaminants from producing defects in the extruded products and/or damaging the die.
- Electrical heater bands and fans that surround the barrel help bring the extruder to operating temperature during start-up and maintain barrel temperature during operation. Several heater bands, one fan, a temperature sensing element called a thermocouple, and a temperature controller are grouped into one temperature zone.
- A rupture disk is located at the end of the extruder barrel, just before the breaker plate. This rupture disk opens when the pressure in the extruder exceeds the rated pressure of this rupture disk. Typical pressures are 5,000, 7,500, and 10,000 psi. Rupture disks are safety devices that are required by law for all extruders used in the United States.

During extrusion, solid resin in the form of pellets or powder is fed from the hopper, through the feed port, into the feed throat of the extruder. The solid resin falls onto the rotating screw and is packed into a solid bed in the first section of the screw called the feed zone. The solid bed is melted as it travels through the middle section of the screw. The melt is mixed, and pressure is generated in the final section of the screw. Although the heater bands and cooling fans maintain the barrel at a set temperature profile, conduction from the barrel walls provides only 10%−30% of the energy required to melt the resin. The remainder of the energy is generated from the frictional heat generated by the mechanical motion of the screw.

Extruder screws are designed to accommodate this pattern of packing, melting, and pressure generation. As shown in Fig. 9.2 the outside diameter of the screw, which is measured at the top of the screw flights, remains constant [1]. The root diameter of the screw, however, changes. In the feed zone, the root diameter is small so that the large channel depth can accommodate the packed solid resin particles. The root diameter of the transition or compression zone increases with the distance from the feed zone. This change in channel depth forces the solid into better contact with barrel wall, thereby promoting better melting. It also compresses the molten polymer in the screw channels. The root diameter becomes constant again

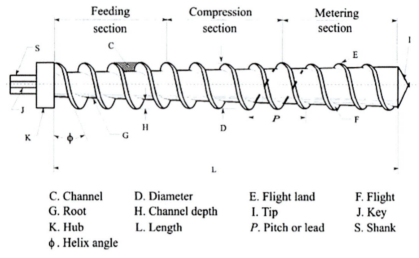

C. Channel D. Diameter E. Flight land F. Flight
G. Root H. Channel depth I. Tip J. Key
K. Hub L. Length P. Pitch or lead S. Shank
φ. Helix angle

Figure 9.2 General-purpose extruder screw [1].

in the metering zone, but the channel depth is very small. This geometry facilitates pressure generation and helps maintain the temperature of the polymer melt. The compression ratio (ratio of the channel depth in the feed and metering zones) and length of the transition zone significantly affect the melting in the single-screw extruders.

9.2.2 Tubing extrusion

Fig. 9.3 shows schematics of an extrusion line to produce tubing. Pressure generated in the extruder forces the melt through the breaker plate, die adaptor, and die. The die forms the melt into the desired shape. Downstream equipment, such as water bath, cools the melt, and the puller draws the extrudate away from the die and through the water bath [1].

9.2.3 Film extrusion

Film extrusion is one of the major processes used for manufacturing plastic films. In this process, the extruder pumps molten resin through a flat film or

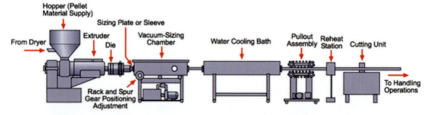

Figure 9.3 Schematics of an extrusion line to produce tubing [1].

sheet die. The melt leaves the die in the form of a wide film or sheet. This is typically fed into a chill roll assembly that stretches and cools the film. After the film passes through the rolls, the film is wound up on a roll.

A film or sheet extrusion line consists of an extruder, film or sheet die, cooling system, take-off system, wind-up system, and auxiliary equipment, such as film gauging systems, surface treatment, and slitters. Single-screw extruders with relatively long barrels (L/D = 27 to 33:1) are used for most resins. Film is usually extruded down onto a chill roll assembly. The film is cooled and drawn by two or more water-cooled rolls, while the surface of chill rolls imparts a finish to the film.

Film extrusion dies are wide flat dies consisting of two pieces: a manifold and the die lip. The manifold distributes the melt across the width of the die, whereas the die lip controls molten film thickness.

Three basic manifold designs that are used in flat film dies include the following:
- T-design manifold
- Coathanger manifold
- Fishtail manifold

The T-design is simple and easy to manufacture. Although it produces a nonuniform pressure drop across the die, and thus, causes nonuniform flow through the die lip, the distribution of the melt does not produce die distortion or clam shelling. This design is not suitable for high-viscosity or easily degradable melts but can be used for extrusion coating [2,4].

In the coathanger manifold, a channel distributes the melt whereas flow is restricted in the preland. While the shape of the manifold channel compensates for the pressure drop, the placement of the die bolts permits die distortion that varies with melt viscosity and polymer flow rate [2,4].

With a fishtail manifold, the entire land area, rather than a flow channel, changes to adjust the pressure drop. This gives better melt distribution, but the die is massive and contains a large mass of polymer. As a result, the fishtail manifold can create temperature nonuniformities and degradation problems [2,5].

9.3 Injection molding process

Injection molding is the most widely used process for producing medical devices, especially for those that are complex in shape and require high-dimensional precision. A conventional injection molding machine is shown in Fig. 9.4 [6].

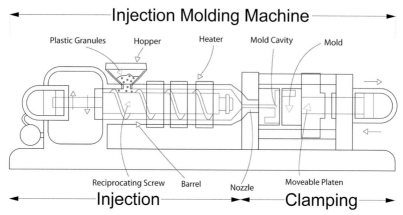

Figure 9.4 Injection molding machine [6].

Plastic pellets are fed into the barrel of the injection molding machine where shear heat is generated by the rotating screw and external heat provided by electric heaters around the barrel melt the plastic, making it ready to be processed. As the screw rotates, it pushes the required amount of plastic for the shot to the front of the barrel. This plastic is injected into the mold by the forward movement of the screw. Once the part is cooled in the mold, it is ready to be ejected out of the mold.

The injection molding screw is responsible for achieving homogeneous melt. The screw provides shear heat to assist in the melting process along with the required mixing and homogenizing of the melt. It also helps in accurately measuring the volume of the shot to injected into the mold. General-purpose screw is the most common injection molding screw and is shown in Fig. 9.5 [6].

Injection molding screws are similar to extrusion screws, which have three zones, i.e., feed zone, transition or compression zone, and metering zone. In the feed zone, the root diameter is small so that the large channel depth can accommodate the packed solid resin particles. The root diameter of the transition or compression zone increases with the distance from the feed zone. This change in channel depth forces the solid into better contact

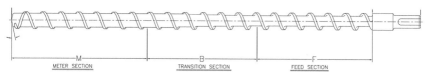

Figure 9.5 A general-purpose injection molding screw [6].

with barrel wall, thereby promoting better melting. It also compresses the molten polymer in the screw channels. The root diameter becomes constant again in the metering zone, but the channel depth is very small [6].

9.4 Catheter manufacturing process

The performance characteristics necessary for catheters to reach smaller distal areas within the body include pushability, torque transmission, and flexibility. Catheters are inserted into a vascular passage from a peripheral location, such as the femoral artery. Three primary variables affect the performance characteristics of a device: materials, design, and process. Outside dimensions are constrained by the size and configuration of the vascular passageway through which it is inserted. Inside dimensions are often constrained by the necessary working channels through which fluids, gases, or devices must pass for diagnosis or therapeutic intervention [7].

Catheters can be either single lumen or multilumen, which are used for transporting liquids, gases, or surgical devices during a medical procedure. For example, single-lumen catheters are most commonly used in IV, urological, and drainage catheters. Two- and three-lumen catheters are often used in peripherally inserted central catheter (PICC) lines. Percutaneous transluminal coronary angioplasty (PTCA) catheters with two-lumen configurations have a small round lumen for guidewires and a large crescent lumen for balloon inflation [7].

Catheter manufacturing consists of the following process steps:
- Step 1—Extrusion of catheter shaft and balloon tubing
- Step 2—Secondary operations

9.4.1 Extrusion process

Catheter extrusion equipment consists of the following:
(a) Resin dryer
   Resins such as PET and polyamides require predrying before they could be used in catheter manufacturing. It is important to follow manufacturer's drying guidelines. Dryers can be standalone or connected directly to the extruder hopper for on-demand drying.
(b) Extruder assembly
   Extruder assembly consists of the following:
   - Extruder
   - Screw
   - Die assembly

(i) Extruder

Single-screw extruder is the common choice within medical device industry. Extruder is sized based on the need of the customer.

Some of the key considerations that need to be considered are as follows:
- What type of tubing being extruded, i.e., single-lumen versus multilumen?
- What size of tubing being extruded, i.e., small versus large?
- What type of material will be used?
- Future upgrades

(ii) Screw

Screw is selected based on the type of material being extruded. Typically, it is recommended to use barrier screw designs for homogeneous mixing [7]. Screw flights can either be ion-nitride or chrome-hardened. However, chrome is the preferred choice when it comes to selecting the hardness of the screw within the medical device industry. This is because chrome-hardened screws are generally less susceptible to chipping over extended use and do not contaminate the product. Depending on the extruder specification, correct screw L/D can be selected.

(iii) Die assembly

Die assembly consists of the following components:
- Breaker plate
- Screen pack
- Inlet and outlet die adapters
- Crosshead assembly
- Melt pump
- Die and mandrel

Screen pack is a group of metal-wire screens placed at the end of the screw and supported by the breaker plate. Breaker plate is a circular disc with holes or slots. Both the screen pack and the breaker plate together provide a seal between the extruder and the die and prevent contaminant from passing through. Screens also help in improving mixing and protect melt pump from being damaged by contaminants [8].

A rupture disc is a hollow, threaded plug with a thin metal diaphragm welded to the plug housing. The diaphragm is exposed to the melt on one side. If the melt pressure exceeds the strength of the diaphragm, it will burst and reduce the extrusion pressure by venting material from the machine. Such a plug is generally located between the screw tip and the breaker plate [9].

The extruder head is hinged for access so that the screw and the breaker plate can be removed for cleaning when required. When closed, the head must be sealed against the barrel so that material cannot escape, even at maximum extrusion pressure.

One end of the heated inlet adapter holds the breaker plate assembly, which is attached to breaker plate and screen pack clamp assembly [8]. The other end of the inlet adapter is attached to the melt pump, which is then attached to the outlet adapter on the other end. The main function of melt pump is to build pressure and maintain precise flow of material to the die.

Crosshead assembly consists of a crosshead and a flow distributor that provides balanced flow of the extrudate, which is distributed evenly over the tube extrusion. The flow distribution remains uniform even with changes in the extruder speed. The crosshead assembly is attached to the outer adapter. Place die holder assembly on the downstream end of the crosshead assembly. Die holder assembly consists of die holder, die retainer, and die adjustment bolts [8].

At this point, mandrel is inserted through the crosshead from the opening in the back until the mandrel is visible on the downstream end of the assembly. Tip retainer is then threaded to close the opening in the back. Die is placed in the die holder and thread die nut onto the die holder until the die is flush with the end of the mandrel tip. A servo pressure system is connected into the back of the mandrel using air pressure adapter fittings and will supply air to all lumens [8].

(c) Vacuum cooling and sizing tank

Extrudate exiting the die in the form of strand passes through vacuum cooling and sizing tank. The primary function of the tank is to cool the polymer and also maintain size and concentricity.

(d) UV treatment

UV lamp is placed at the end of the vacuum tank, and its function is to UV treat endotoxins that may have transferred onto the tubing (catheter shaft and balloon tubing) from the water in the tank.

(e) Laser and ultrasonic gages

Ultrasonic gage assembly is placed after the UV treatment unit to measure wall thickness of the tubing. Tubing then passes through laser gage system, which measures the outer diameter of the tubing. Based on the thickness and outer diameter data, the software calculates internal diameter of the tubing. Gages transmit the data directly to the measurement software, which then sends signal for the tubing to be rejected or accepted.

Tubing then proceeds to the cutter assembly where the tubing is cut into predefined length and transported through a conveyor belt. Based on the output from the measurement software, the tubing either gets blown into the rejection bin or drops into accepted bin.

9.4.2 Secondary operations

Outside dimensions are constrained by the size and configuration of the vascular passageway through which it is inserted. Inside dimensions are often constrained by the necessary working channels through which fluids, gases, or devices must pass for diagnosis or therapeutic intervention. For these reasons, catheter shaft requires secondary operations to convert into final product.

Below are series of steps:
1. Catheter shaft drawdown
2. RO marker swaging
3. Distal skiving
4. Balloon proximal thermal welding
5. Balloon distal thermal welding
6. Proximal skiving
7. Proximal fuse
8. Manifold bonding
9. Balloon pleating and folding

9.4.2.1 Catheter shaft drawdown

The purpose of this step is to reduce the diameter of the distal end of the catheter. Catheter shaft drawdown process comprises three steps:
- Prestretching
- Drawdown
- Normalization

In the prestretching step, beading is inserted in the distal end of the shaft and the shaft is prestretched to a distance at a set speed. These prestretched shafts are then drawn down under a set temperature and drawdown speed. The drawdown shafts are now ready to be normalized. PTFE-coated mandrels are inserted in the guidewire lumen from the proximal end and transferred into oven which is set at the desired temperature and time. At this point, catheter shafts are ready for the next step.

9.4.2.2 RO marker swaging
RO marker swaging of the catheter shaft is done so that the location of a catheter tip under fluoroscopy can be identified. Platinum iridium metals are the typically bands used in medical industry as they are easily detectable.

The process involves positioning the bands, precrimping the bands into triangular or hexagonal shape to prevent them moving on the tube, and swaging the bands. Swaging is performed on the distal end of the shaft using a swaging machine with a special die that radially hammers the band many times to reduce the diameter and make the band round and smooth. Precrimping prevents the band moving along the tube during swaging [9,10].

9.4.2.3 Proximal and distal skiving
Skiving is performed on the distal end of the shaft where a slice is cut along the long of axis of the tube leaving a thin layer of tubing remaining after completion. It is typically done in applications where a hole in the tubing is needed after the tubing has been joined with another component such as infusion devices, irrigation, and drainage [11].

9.4.2.4 Balloon proximal and distal thermal welding
Balloon thermal welding process is to thermally weld balloon tubing to the catheter shaft using a thermal welder. The first step is to have Teflon beading inserted into the guidewire lumen on the welding end, i.e., distal end for distal welding and proximal for proximal welding. In the next step, the balloon is fully stretched, and the heat shrink is placed over the shaft end (distal end if it is distal welding and proximal end if it is proximal welding) assembly and placed into balloon distal welding equipment. Once the welding operation has concluded, heat shrink from the catheter is peeled off toward the balloon.

9.4.2.5 Proximal fuse
In the proximal fuse process, the proximal end of the shaft with mandrel in guidewire lumen is inserted into preheated mold. Pressure is applied onto the shaft for a specified amount of time making sure that the proximal

end stays firmly inside the mold. Once the heating cycle is completed, the mold is then air cooled. After which, the shaft and mandrel can be taken out of the mold for inspection.

9.4.2.6 Manifold bonding

Manifold bonding is an adhesive bonding of catheter shaft and strain relief to a manifold using a bonding equipment. In this process, a mandrel is inserted in the proximal end of the catheter shaft and the manifold area next to skive holes and the proximal fuse area of the catheter shaft is primed. Once the primer has dried, the strain relief is placed over the proximal end of the mandrel. The distal end of the catheter is held in place using clip on one end, and the manifold is held in a fixture on the other end. The proximal end of the shaft is completely inserted into the manifold, and the complete sample is placed in the bonder for bonding.

9.4.2.7 Balloon pleating and folding

The first step of balloon wrapping is the pleating step. In this step, the inflated balloon is inserted into the pleating station. The pleating dies close around the balloon, forming pleats while the balloon is still inflated. In most applications, the pleating dies are heated. When the vacuum delay time is reached, a vacuum is applied in the balloon. This vacuum ensures that the pleated shape will be retained for the next step, compression [12].

The second step of balloon wrapping is the compression step. In this step, the evacuated balloon is inserted into the compression station. The compression dies close around the pleated balloon, while vacuum is still applied. In most applications, the compression dies are heated. The compression dies can be closed to reach either a force or diameter. When the compression dwell time is complete, the station will open, and the operator can now put a sheath on the balloon while vacuum is still applied.

9.5 Other secondary operations

9.5.1 Laser marking

Identification and traceability are two key elements of medical device industry. Government has set guidelines that require medical device manufacturers to implement use of unique device identification (UDI) on all medical devices, implants, tools, and instruments. This can be achieved by laser marking onto the devices.

Laser marking is a preferred and noncontact form of engraving for product identification on medical devices that offer consistent high-quality laser marks at high processing speeds while eliminating any potential damage or stress to the parts being marked. These markings are also corrosion resistant and can withstand sterilization processes such as passivation, centrifugation, and autoclaving [13].

9.5.2 Lubrication

Lubrication refers to the process of reducing friction and facilitate smooth movement between two surfaces or components by applying friction-reducing substance known as lubricant. As like in any other industry, medical devices are no exception. When two components experience excessive or unexpected friction, it leads to device components to wear out and eventually cause device performance to decline over time. To protect components against friction and wear, friction-reducing substance or film is applied between two surfaces or components [14].

9.5.3 Siliconization

Siliconization is the process of treating or coating the surface between the two components with dimethyldichlorosilane to reduce friction. Typical examples include needles, syringes, guidewires, and catheters. Therefore, it is important to understand the surface onto which it is going to be applied, i.e., silicone, metal, glass, and plastics. Each material may have different characteristics that can pose unique lubrication challenges [15].

There are four types of material surfaces:
- Silicone
- Metal
- Glass
- Plastics

9.5.3.1 Silicone

Silicone is an elastomer that is often used in components within the medical device industry primarily because of its biocompatibility and superior chemical properties. However, it exhibits high coefficient of friction and becomes tacky when it is cured and causes problems when it is injection molded, or extruded. There are factors that need to be understood when selecting silicone [15]:
- Dispersion
- Surface interaction

- Material viscosity
- Processing time
- Moisture sensitivity

Dispersion
It is important to consider dispersion when selecting silicone. There is an advantage of using high-molecular-weight silicone for devices that are intended for multiple uses. During the lubricant coating process, the solvent flashes off and results in strong adhesion.

Surface interaction
It is important to consider the impact of surface interaction between lubricant and silicone. Difference in chemistries between the part and the lubricant itself should be accounted for part molded from silicone. This is important because the lubricant is capable of diffusing into chemically similar material leading molded component to swell. This causes the fluid to deplete from the surface resulting in reduction or elimination of the lubricating effect.

Material viscosity
Material viscosity is another factor that can have significant implications. It is inversely proportional to diffusion into the substrate. The higher the viscosity of the lubricant, the lower the chance of migration. Therefore, a higher-viscosity fluid for longer lubrication periods is considered.

Processing time
Use of self-lubricating silicone elastomers can reduce or eliminate a number of processing steps. For example, adding a lubricant, coating, or grease to the surface of a component or device will not be required since lubricity is inherent property of the silicone elastomer.

Moisture sensitivity
It is critical to evaluate moisture sensitivity since silicone can absorb moisture. Consequently, if adjustments are performed to optimize viscosity or solids content, they should take place in a moisture-free environment.

9.5.3.2 Metals
Metal surfaces and edges of hypodermic and suture needles, scalpels, or other cutting edges have an inherently high surface friction. This can make patient

uncomfortable during penetration in human tissue. This can be resolved by appropriately designing the component. In addition, there are other factors that need to be looked when selecting metals [15]. They are as follows:
- Surface interaction
- Formulation

Surface interaction
There are two main actions that need to be considered—(1) penetration frequency and (2) lubricant longevity. To minimize the effects of surface friction, metal can be coated with silicone lubricants to lower the coefficient of friction without compromising penetration or cutting efficacy.

Formulation
Consider dispersion and bonding behavior. For metal components that require lubricity, it is important to consider dispersion and bonding characteristics. Dispersed silicone formulations minimally bond to the metal substrates they coat, making them ideal to lubricate needles. For example, polydimethlysiloxane (PDMS) fluid is typically considered for this substrate.

9.5.3.3 Glass
Silicone fluids have a silicon-oxygen chemical structure similar to glass, quartz, and sand. Consequently, they tend to bond very well with glass. Cross-linking to enhance bonding over the glass substrate may be achieved by heating the silicone beyond its operating temperature. Factors that need to be considered when selecting glass as a substrate [15]:
- Formulation
- Curing

Formulation
To reduce drag forces in glass-prefilled syringes, use of hydrophobic lubricant would be an ideal choice. For example, the insides of the glass syringes can be coated with a polydimethylsiloxane (PDMS) oil.

Curing
Consider using high-temperature heating to activate cross-linking characteristics of PDMS. PDMS by itself is nonfunctional and does not cure. However, by exposing glass syringes to extremely high temperatures will activate polymer cross-linking.

9.5.3.4 Plastics

There are a wide range of plastics that are used in medical devices, for example, valves, syringes, and IVDs. All of these can benefit from silicone lubricants. However, there are some considerations that need to be taken when selecting plastics as a substrate [15]:

Formulation

For applications that involve slide and glide function, a very-high-viscosity grease can be considered such as silicone grease providing lubricity and minimizing migration when applied to a plastic surface.

References

[1] S. Ashter, Chapter 10: other processing approaches, in thermoforming of single and multilayer laminates, Elsevier, 2013, pp. 230—260.
[2] J. Vlachopoulos, N.D. Polychronopoulos, S. Tanifuji, P.J. Müller, Chapter 4: flat film and sheet dies, in: O.S. Carneiro, M. Nobrega (Eds.), Design of Extrusion Forming Tools, Smithers Rapra, London, UK, 2012, pp. 113—140.
[3] J. Ivey, in: J. Vlachopoulos, J. Wagner (Eds.), The SPE Guide on Extrusion Technology and Troubleshooting, Society of Plastics Engineers, Brookfield, CT, USA, 2001, p. 12.1.
[4] D.R. Garton, in: T.I. Butler, E.W. Veazey (Eds.), Film Extrusion Manual, TAPPI, Atlanta, GA, 1992, p. 231.
[5] W. Michaeli, Extrusion Dies for Plastics and Rubber, second ed., Hanser Publishers, Munich, 1992.
[6] Brendan Rockey, Injection Molding Machine, Wikimedia Commons. https://commons.wikimedia.org/wiki/File:Injection_moulding.png.
[7] B. Flagg, New Extrusion Techniques Advance Catheter Design, Medical Device and Diagnostic Industry, 2013. https://www.mddionline.com/news/new-extrusion-techniques-advance-catheter-design.
[8] J. Goff, T. Whelan, Section 4: The Screw and Barrel System, Dynisco Extrusion Process Handbook, second ed., in: D. Delaney (Ed.).
[9] Marker Band Swagers, Blockwise. http://blockwise.com/swagers/.
[10] D. Sanchez, The Benefits of Embedding Marker Bands vs Swaging, Medical Design Briefs, 2020. https://www.medicaldesignbriefs.com/component/content/article/mdb/tech-briefs/37126.
[11] Skived Tubing, Duke Extrusion. https://www.dukeextrusion.com/tubing-options/skived-tubing.
[12] Balloon Wrappers, Blockwise. http://blockwise.com/balloon-wrappers/.
[13] Laser for marking medical devices, Laserstar Technologies. https://www.laserstar.net/en/industries/medical/marking-and-engraving/.
[14] Lubricants, Nye Lubricants. https://www.nyelubricants.com/medical.
[15] J. Chandler, Key Factors for Choosing Silicone Solutions in Medical Device Lubrication, Medical Design Brief, 2018. https://www.medicaldesignbriefs.com/component/content/article/mdb/features/technology-leaders/28752.

CHAPTER TEN

Therapeutic applications of medical devices

Contents

10.1	Biosurgery	164
	10.1.1 Reinforced bioscaffolds	165
	10.1.1.1 Tela Bio OVITEX	*165*
	10.1.1.2 Tela Bio OVITEX PRS	*165*
	10.1.2 AROA Bio endoform restorative bioscaffolds	165
	10.1.3 Engineered extracellular matrix	166
	10.1.3.1 AROA bio myriad matrix	*166*
	10.1.3.2 AROA bio myriad morcells	*166*
	10.1.4 Tissue products	166
	10.1.4.1 dCELL technology	*167*
	10.1.5 Hemostats	167
	10.1.5.1 Arista AH absorbable hemostat	*168*
	10.1.5.2 Avitene hemostats	*168*
	10.1.6 Sealants	168
	10.1.6.1 Progel pleural thoracic sealant	*168*
	10.1.6.2 Tridyne vascular sealant	*169*
	10.1.6.3 Vistaseal fibrin sealant	*169*
10.2	Cardiovascular heart rhythm	169
	10.2.1 Automated external defibrillators (AEDs)	169
	10.2.1.1 Cardiac Science Corporation	*170*
	10.2.1.2 Zoll Medical Corporation	*170*
	10.2.1.3 Defibtech LLC	*172*
	10.2.1.4 Philips medical systems	*172*
	10.2.2 Cardiac ablation catheter	173
	10.2.2.1 Medtronic's DiamondTemp ablation system with RealTemp	*174*
	10.2.2.2 Medtronic's family of cardiac cryoablation catheters	*175*
	10.2.3 Cardiac pacemakers	175
	10.2.4 Implantable cardioverter defibrillators (ICDs)	175
10.3	Vascular surgery	178
	10.3.1 Vascular grafts	178
	10.3.1.1 Flixene AV Access Grafts	*178*
	10.3.1.2 Advanta VXT vascular graft	*181*
	10.3.2 Vascular patches	182
	10.3.3 Tunneling solutions	182
	10.3.3.1 Vascular graft tunneling instrumentation	*182*
	10.3.3.2 SLIDER graft deployment system	*184*

Applications of Polymers and Plastics in Medical Devices
ISBN: 978-0-12-820980-6
https://doi.org/10.1016/B978-0-12-820980-6.00006-0

© 2022 Elsevier Inc.
All rights reserved.

	10.3.4 Jetstream atherectomy system	185
10.4	Interventional cardiology	185
	10.4.1 WATCHMAN left atrial appendage closure device	185
	10.4.2 SYNERGY bioabsorbable polymer stent	187
	10.4.3 XIENCE Sierra everolimus eluting coronary stent system	188
	10.4.4 NC Trek coronary dilation catheter	188
	10.4.5 Trek and Mini Trek coronary dilation catheter	188
	10.4.6 XIENCE Alpine everolimus eluting coronary stent system	190
10.5	Thoracic drainage systems	191
10.6	Prosthetic implants	193
	10.6.1 Stryker corporation	193
	10.6.2 Johnson and Johnson	194
	10.6.3 Otto Bock	194
	10.6.3.1 Prosthetic knee solution	*195*
	10.6.3.2 Prosthetic foot solution	*196*
	10.6.3.3 Prosthetic socket solution	*196*
	10.6.3.4 Prosthetic hip solutions	*197*
	10.6.4 Smith and Nephew	197
10.7	In vitro diagnostics	197
	10.7.1 Abbott laboratories	198
	10.7.2 Bio-Rad laboratories Inc.	198
	10.7.3 Danaher corporation	199
	10.7.3.1 Automate 2500 family	*200*
	10.7.3.2 Power link workcell	*200*
	10.7.3.3 Power express laboratory automation system	*200*
	10.7.3.4 DxA5000 total laboratory automation system	*200*
	10.7.4 Hologic Inc.	200
	10.7.4.1 Panther system	*201*
	10.7.4.2 Panther fusion system	*201*
	10.7.4.3 Panther plus	*202*
	10.7.4.4 Panther link	*203*
References		203

10.1 Biosurgery

Biosurgery is a type of surgical method that involves use of natural and synthetic surgical products to control bleeding. The success of surgical procedures depends on effective control of bleeding, even for minimally invasive procedures. Biosurgery products are divided into the following categories:
- Reinforced bioscaffolds
- Restorative bioscaffolds

- Engineered extracellular matrix (ECM)
- Tissue products
- Hemostats
- Sealants

10.1.1 Reinforced bioscaffolds
10.1.1.1 Tela Bio OVITEX
Tela Bio OVITEX is a reinforced bioscaffold with a tissue matrix designed for hernia repair and abdominal wall reconstruction. It comprises polymer fiber looped through layers of biologic material, one on each side to create a lockstitch pattern. The biologic material is derived from ovine rumen that reduces foreign body response and promote hernia repair. Unique permeable design facilitates rapid fluid transfer and movement of cells through the product. Interwoven polymer provides strength and improved handling while the lockstitch pattern provides additional reinforcement that improves handling and load sharing capability for supporting natural abdominal wall function that prevents unraveling when cut [1].

10.1.1.2 Tela Bio OVITEX PRS
Tela Bio OVITEX PRS is a reinforced bioscaffold with a tissue matrix designed for soft tissue repair in plastic and reconstructive surgery. Similar to OVITEX, OVITEX PRS comprises layers of ovine rumen that is interwoven with polymer fibers using patented corner-lock diamond embroidery construction. The corner-lock diamond embroidery pattern allows the product to stretch while maintaining its shape. In order to have improved fluid management, tissue integration and directional flexibility, layers of ovine lumen are carefully designed to incorporate micropores [2].

10.1.2 AROA Bio endoform restorative bioscaffolds
AROA Bio has developed restorative bioscaffolds using AROA ECM technology that contains 0.3% ionic silver and uses 150 ECM proteins enabling it to interact with patient's cells throughout all phases of [3].

There are two main types of restorative bioscaffolds:
- Endoform antimicrobial restorative bioscaffold
- Endoform natural restorative bioscaffold
  Advantages of restorative bioscaffolds include:
- They help identify presence or absence of proteases and restores protease balance.

- They facilitate rapid establishment of capillary network by supporting migrating endothelial cells.
- They establish new vascular channels and increase blood supply to help build new tissues.

10.1.3 Engineered extracellular matrix

10.1.3.1 AROA bio myriad matrix

AROA Bio Myriad Matrix is an engineered ECM for soft tissue repair, reinforcement, and complex wound applications. They are designed to help maximize tissue repair in a wide range of surgical applications [4].

Advantages of Myriad Matrix include [4]:
- It is strong, soft, drapable, and conforming.
- It rehydrates quickly and is easy to cut, suture, or staple.
- It helps in routine repair and reinforcement procedures.
- Fast absorption of blood and cells accelerates host cell infiltration.
- It contains 150 ECM proteins that help aid in the healing process.
- It supports blood vessel development imparted by structural effect of vascular channels on endothelial cells known as angioconduction.

10.1.3.2 AROA bio myriad morcells

AROA Bio Myriad Morcells is a powdered version of AROA Bio Myriad Matrix that is used for soft tissue repair and complex wound applications.

Myriad Morcells^{TM} delivers extracellular matrix (ECM) proteins that aids in the healing process. Their powdered format increases the extracellular matrix surface area to maximize the delivery [5].

Myriad Morcells offers similar advantages as Myriad Matrix [5]:
- It contains 150 ECM proteins that help aid in the healing process.
- Fast absorption of blood and cells accelerates host cell infiltration.
- It supports blood vessel development imparted by structural effect of vascular channels on endothelial cells.
- It acts as a synergy to further accelerate cell infiltration into the Myriad Matrix.

10.1.4 Tissue products

Unlike reinforced and restorative bioscaffolds that use 150 ECM proteins enabling it to interact with patient's cells throughout all phases of healing, there are tissue products that use specialized technology to produce decellularized tissue, which serves as replacement for the damaged ECM [6].

10.1.4.1 dCELL technology
Tissue Regenix Biosurgery offers two tissue products produced using patented dCELL technology [6]:
- DermaPure
- SurgiPure

10.1.4.1.1 DermaPure
DermaPure is produced using patented dCELL technology and is intended to provide reinforcement, repair, or replacement of damaged tissue. Following is the list of applications where DermaPure is used [7]:
- Traumatic injuries
- Cancerous lesion resection
- Head/neck reconstruction
- Surgical dehiscence
- Pilonidal cysts
- Hidradenitis suppurativa
- Necrotizing fasciitis
- Diabetic limb salvage
- Chronic wounds

10.1.4.1.2 SurgiPure
SurgiPure is a reconstructive tissue matrix that is produced using dCELL technology and is intended to use as a reinforcement to support soft tissue at its weak point and for the surgical repair of damaged or ruptured soft tissue membranes. It consists of decellularized collagen ECM that allows repair by repopulation and integration of native cells. SurgiPure is biomechanically strong and biocompatible and allows incorporation of recipient cells [8].

10.1.5 Hemostats
Hemostats are surgical tool used in surgical procedures to control bleeding. They have three main functions [9]:
- Hemostats are used to clamp small blood vessels for hemorrhage control.
- Hemostats are used to grasp and secure superficial fascia during undermining and debriding wounds.
- Hemostats are used to explore and visualize the deeper areas of a wound.
  Becton Dickinson (BD) is one of the leaders in hemostats and offers a range of products that include [10−14]:
- Arista AH absorbable hemostat
- Avitene microfibrillar collagen hemostat

- Avitene sheet
- Avitene Ultrafoam collagen sponge
- EndoAvitene applicators
- SyringeAvitene Applicators

10.1.5.1 Arista AH absorbable hemostat
BD's Arista AH absorbable hemostat is an absorbable surgical hemostat in powder form that is derived from purified plant starch [10]. It is used in situations when other conventional methods are unable to control capillary, venous, or arteriolar bleeding [10].

10.1.5.2 Avitene hemostats
BD's Avitene hemostats are collagen-based hemostats that accelerate clot formation. They are known to increase platelet aggregation and the release of proteins to form fibrin, resulting in hemostasis. There are five types of Avitene hemostats available [11–14]:
- Avitene microfibrillar collagen hemostat
- Avitene sheets
- Avitene Ultrafoam collagen sponge
- EndoAvitene applicators
- SyringeAvitene applicators

10.1.6 Sealants
Sealants are used to seal air leaks during surgeries. BD is the major supplier of sealants and has two products on the market that provide effective solution to air leaks during thoracic and vascular surgeries [15,16]. They are as follows:
- Progel pleural thoracic sealant
- Tridyne vascular sealant

Johnson and Johnson (J&J) is another player on the sealant market. The Vistaseal Fibrin sealant triggers clot formation in a patient irrespective of their coagulation profile and successfully seals the leak.

10.1.6.1 Progel pleural thoracic sealant
BD's Progel pleural thoracic sealant is an FDA-approved sealant that is known to seal air leaks during thoracic surgery [15]. Progel allows the lung to expand and control during respiration due to its unique combination of strength, flexibility, and adherence.

10.1.6.2 Tridyne vascular sealant

BD's Tridyne vascular sealant seals air leaks during cardiovascular, cardiothoracic, and vascular surgeries reinforcing aortic anastomoses and control bleeding. The sealant consists of BD's proprietary formulation of polyethylene glycol (PEG) and human serum albumin, which forms a strong and flexible seal, even in anticoagulated patients [16].

10.1.6.3 Vistaseal fibrin sealant

J&J's Vistaseal fibrin sealant contains a combination of clotting proteins found in human plasma, fibrinogen, and thrombin, which triggers clot formation in a patient irrespective of their coagulation profile. The sealant provides sustained hemostasis, faster preparation time and flexibility, precision, and broad working temperature with no additional warming required [17].

10.2 Cardiovascular heart rhythm

Heart rhythms are electrical signals or impulses that coordinate heartbeats. When these electrical impulses do not work properly, they can cause variation in heartbeat. This situation is termed as heart arrhythmia. In most cases, heart arrhythmias may be harmless; however, there are some situations where heart arrhythmias may be bothersome and even be life-threatening [18–24].

There are five types of medical devices in the market that have been approved by FDA to treat heart arrhythmia:
- Automated external defibrillators (AEDs)
- Cardiac ablation catheters
- Cardiac pacemakers
- Implantable cardioverter defibrillators (ICDs)
- Ventricular assist devices (VADs)

10.2.1 Automated external defibrillators (AEDs)

Automated external defibrillator (AED) is a medical device that analyzes the heart rhythm of a person and delivers an electric shock to restore the uncoordinated heart rhythm to normal. This uncoordinated heart rhythm is called ventricular fibrillation, which causes the person to become unresponsive leading to sudden cardiac arrest.

AEDs are FDA-regulated and the following are approved manufacturers:
1. Cardiac Science Corporation
2. Defibtech LLC

3. Philips Medical Systems
4. Zoll Medical Corporation

10.2.1.1 Cardiac Science Corporation

Cardiac Science Corporation, one of the leading manufacturers of Automatic External Defibrillator (AED) is part of Zoll Medical Corporation's Resuscitation division. It has four models on the market:
- Powerheart AED G3, G3 Plus, and G5
- Powerheart AED G3 Pro

Powerheart AED G3, G3 Plus, G5, and G3 Pro are portable, battery-operated, self-testing unit AEDs that are capable of diagnosing and treating patients with life-threatening heart arrhythmias [18,19].

Powerheart AED G3, G3 Plus, G5 (Fig. 10.1), and Powerheart G3 Pro (Fig. 10.2) use two multifunction defibrillation electrodes that are placed on the patient's chest, to acquire a patient's electrocardiogram (ECG). If a defibrillation shock is required, the device will deliver a biphasic, impedance compensating, and truncated exponential shock waveform to the patient through the multifunction defibrillator electrodes [18,19].

10.2.1.2 Zoll Medical Corporation

Zoll Medical Corporation manufactures Automatic External Defibrillators (AEDs) that are intended for trained medical personnel to treat patients with sudden cardiac arrest. Some of the types includes X Series®, R Series®, Propaq® MD, AED Pro®, and AED 3 BLS® as shown in Fig. 3 [20]. These models use two multifunction defibrillation electrodes to acquire a patient's electrocardiogram. If a defibrillation shock is required, the device will deliver an escalating, impedance-compensating, rectilinear biphasic shock waveform through the multifunction defibrillator electrodes.

Powerheart® AED G5   Powerheart® AED G3 and AED G3 Plus

Figure 10.1 Cardiac Science's Powerheart AED G3, G3 Plus, and G5 [18]. *Source: Food and Drug Administration, Republished on 01/08/2022.*

Figure 10.2 Cardiac Science's Powerheart AED G3 Pro [19]. *Source: Food and Drug Administration, Republished on 01/08/2022.*

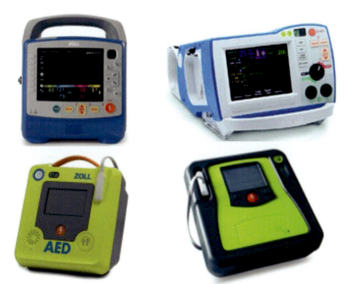

Figure 10.3 Zoll medical corporation's automatic external defibrillators [20]. *Source: Food & Drug Administration, Republished on 01/08/2022*

Some of the other capabilities include:
- Electrocardiogram monitoring
- Cardiopulmonary resuscitation monitoring
- External transcutaneous pacing
- Blood pressure monitoring
- Blood oxygen saturation monitoring

- Respiration monitoring
- Carbon dioxide monitoring

10.2.1.3 Defibtech LLC

Lifeline public access defibrillators are family of portable, battery-operated AEDs as shown in Fig. 10.4, which are capable of diagnosing and treating patients with life-threatening heart arrhythmias [21].

The Lifeline AEDs use two multifunction defibrillation electrodes, placed on the patient's chest, to acquire a patient's ECG. The device will provide an audible rhythmic beeping sound to help the user deliver the correct rate of compressions while giving CPR. If a defibrillation shock is required, the device will prompt the user to deliver an electrical shock, through the electrodes [21].

10.2.1.4 Philips medical systems

Philips Medical Systems HeartStart OnSite and HeartStart Home (Fig. 10.5), HeartStart FR3 (Fig. 10.6), and HeartStart FRx (Fig. 10.7) are portable, battery-powered AEDs that are capable of diagnosing and treating patients with life-threatening heart arrhythmias [22–24]. They use two multifunction defibrillation electrodes, placed on the patient's chest, to get a patient's electrical activity of the heart ECG.

Figure 10.4 Defibtech's lifeline AED [21]. *Source: Food and Drug Administration, Republished on 01/08/2022.*

Therapeutic applications of medical devices 173

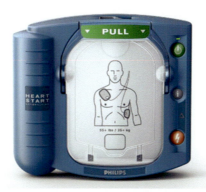

Figure 10.5 Philips HeartStart onsite/home automatic external defibrillator (AED) [22]. *Source: Food and Drug Administration, Republished on 01/08/2022.*

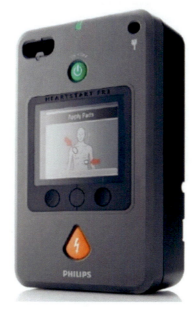

Figure 10.6 Philips HeartStart® FR3 Automatic External Defibrillator [23]. *Source: Food and Drug Administration, Republished on 01/08/2022.*

10.2.2 Cardiac ablation catheter

Cardiac ablation is a minimally invasive procedure to stop heart abnormalities. In this process, a catheter is inserted through the blood vessels to the heart to stop heart abnormality also known as heart arrhythmias. The

Figure 10.7 Philips HeartStart® FR3 Automatic External Defibrillator [24]. *Source: Food and Drug Administration, Republished on 01/08/2022.*

catheter is referred to as cardiac ablation catheter. Some of the leading medical device companies that manufacture ablation catheter includes Biomerics, Boston Scientific, Biosense Webster, and Medtronics.

10.2.2.1 *Medtronic's DiamondTemp ablation system with RealTemp*

Medtronic's DiamondTemp ablation system (DTA) as shown in Fig. 10.8 is the only open-irrigated RF catheter that uses industrial diamonds to optimize power based on the surface temperature of the tissue. Chemical vapor deposition (CVD) of these industrial diamonds acts as heat-shunting material. DTA is safe and effective when compared with contact force-sensing RF. The catheter tip comprises platinum-iridium while the distal, center, and proximal end of the catheter is made from diamond [25].

Figure 10.8 Medtronic's DiamondTemp ablation system with RealTemp [25]. *Source: Image reproduced with permission from Medtronic.*

10.2.2.2 Medtronic's family of cardiac cryoablation catheters
Medtronic's family of cardiac cryoablation catheters includes Arctic Front, Arctic Front Advance, and Arctic Front Advance Pro catheters. Arctic Front catheter as shown in Fig. 10.9 is flexible, over-the-wire balloon catheters used to ablate cardiac tissue. Arctic Front Advance features improved temperature uniformity with EvenCool cryo technology that uses eight jets and enables more contiguous lesions. The next-generation Arctic Front Advance Pro Cryoballoon improves time-to-isolation that can result in improved procedural efficiency [26].

10.2.3 Cardiac pacemakers
Cardiac pacemaker is a medical device that generates electrical impulses delivered by electrodes causing the heart muscle chambers to contract and pump the blood thereby, regulating the function of the electrical conduction system of the heart. Boston Scientific and Medtronics are the two leading manufacturers of pacemakers. Comparison of cardiac pacemakers manufactured by Boston Scientific and Medtronics is shown in Table 10.1 [27,28].

10.2.4 Implantable cardioverter defibrillators (ICDs)
An implantable cardioverter defibrillator (ICD) is a small battery-powered device that is placed in patient's chest to monitor the heart rhythm and

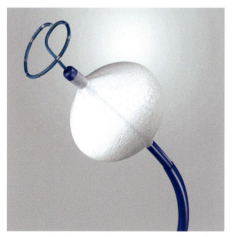

Figure 10.9 Medtronic's Arctic Front cardiac cryoablation catheter [26]. *Source: Image reproduced with permission from Medtronic.*

Table 10.1 Comparison of cardiac pacemakers [27,28].

Boston scientific		Medtronic	
Accolade and Essentio	**Micra AV and Micra VR**		
 Features • RightRate respiration-based pacing • Automatic daily monitoring	Features • Smallest line of pacemakers • Does not leave bump under the skin • No chest scar and require no lead. • Micra is completely self-contained		
Accolade MRI and Essentio MRI	**Azure**		
 Features • Most advanced pacemakers • Can be used as part of the ImageReady	Features • Equipped with BlueSync technology and is compatible with		

- safe and effective scanning in 1.5T and 3T MRI environments
- MyCareLink Heart mobile app
- If Azure detects changes in your heart, it wirelessly and securely transfers your heart device information to your clinic
- Azure pacemaker is safe in the MRI environment

Ingenio and Advantio

Features
- Innovative high-voltage platform
- New features, therapies, and diagnostic options are included

Advisa MRI

Features
- Designed for safe use in the MRI environment
- Available in single and dual chamber options.

Source: Images are reproduced with permission from Boston Scientific and Medtronic.

detect irregular heartbeats. The device continuously monitors the heartbeat and delivers electrical pulses to restore a normal heart rhythm when required.

Medtronics and Boston Scientific are the two main manufacturer of ICDs. Tables 10.2 [29] and 10.3 [30–33] provides information on their ICD products.

10.3 Vascular surgery

Vascular surgery is a surgical subspecialty in which diseases of the vascular system are managed by medical therapy, minimally invasive catheter procedures, and surgical reconstruction. Open surgery techniques and endovascular techniques are used to treat vascular diseases. Some of the available tools include the following:
- Vascular grafts
- Vascular patches
- Tunneling solutions
- Jetstream atherectomy system
- Angiojet thrombectomy system
- Epic vascular self-expanding stent
- Peripheral vascular embolization

10.3.1 Vascular grafts

Vascular grafts are used in a surgical procedure known as vascular bypass that is done to redirect blood flow from one area to another by connecting blood vessels. Vascular grafts are of three types [34]:
- Expanded polytetrafluoroethylene (ePTFE)
- Polyethylene terephthalate (Dacron)
- Polyurethanes

Out of the three vascular grafts, expanded PTFE grafts are widely used in surgical procedures. Getinge is one of the leading manufacturers of ePTFE grafts, and its products are used globally. Getinge's two main ePTFE product lines include the following:
- Flixene AV Access Grafts
- Advanta VXT Vascular Graft

10.3.1.1 Flixene AV Access Grafts

Flixene AV access graft is a three-layered ePTFE graft as shown in Fig. 10.10. The top layer has large pore surface area of 60 μm and is more

Therapeutic applications of medical devices 179

Table 10.2 Medtronic's implantable cardioverter defibrillator (ICD) portfolio [29].

Cobalt XT	
	Features • BlueSync technology enables tablet-based programming and app-based remote monitoring • Device includes an automated antitachycardia pacing algorithm • Approved for 1.5T and 3T MR conditional use
Cobalt	
	Features • BlueSync technology enables tablet-based programming and app-based remote monitoring • Approved for 1.5T and 3T MR conditional use
Crome	
	Features • BlueSync technology enables tablet-based programming and app-based remote monitoring • Approved for 1.5T and 3T MR conditional use
Visia AF and Visia AF MRI	
	Features • Single-chamber ICDs that can detect atrial fibrillation using a traditional lead • Approved for 1.5T and 3T MR conditional use
Evera MRI	
	Features • Approved for full body scanning in the MR environment • Approved for 1.5T and 3T MR conditional use

(*Continued*)

Table 10.2 Medtronic's implantable cardioverter defibrillator (ICD) portfolio [29].—cont'd

Primo MRI		
	Features	
	• Works on SmartShock 2.0 technology and the PhysioCurve design	
	• Approved for 1.5T and 3T MR conditional use	
Evera		
	Features	
	• First ICD device to introduce the PhysioCurve design	

Source: Images are reproduced with permission from Medtronic.

Table 10.3 Boston scientific's implantable cardioverter defibrillator (ICD) portfolio [30—33].

Emblem MRI S-ICD and Emblem S-ICD		
	Features	
	• Subcutaneous implantable cardioverter defibrillators	
	• Continuous monitoring of heart rhythm 24 h a day	
	• Allows MRI screening	
Energen ICD and Punctua ICD	Features	
	• Capable of treating fast and slow heart rhythm	
	• Battery life for Energen ICD lasts for 9—11 months	
S-ICD system		
	Features	
	• Treats sudden cardiac arrest	
	• Battery life for S-ICD system lasts for 5 years	

Source: Images are reproduced with permission from Boston Scientific.

Figure 10.10 Flixene AV access grafts [35]. *Source: Image reproduced with permission from Getinge.*

receptive to tissue growth. The middle layer acts as a reinforcing wrap for additional support. The bottom layer is a small pore layer with an inner graft surface porosity of 20 μm [35].

Flixene grafts come in two different configurations:
- Straight
- Tapered

Straight grafts are often used for applications where bypass is needed and where vessel diameter remains constant from the proximal to the distal anastomosis. Tapered grafts are used where flow dynamics need to be altered or when performing bypass between the two different diameter vessels. Each of these grafts can either be standard wall (SW) or graduated wall (GW) [35].

Some of the other key benefits of Flixene AV grafts include [35]:
- offer a strong and durable cannulation zone engineered to meet the clinical demands for dialysis access;
- ability of Flixene to be safely cannulated within 24–72 h, enabling patients to begin dialysis within days rather than weeks; and
- use of the SLIDER GDS has been associated with reduced complication (seroma and pseudoaneurysm formation) as well as reduced soft tissue trauma occurrence during the tunneling process.

Flixene grafts work with unique SLIDER Graft Deployment System (GDS) that is designed to make tunneling easier than conventional practices, minimize soft tissue trauma, and reduce graft sweating [36].

10.3.1.2 Advanta VXT vascular graft

The Advanta VXT vascular graft is a synthetic vascular graft made from ePTFE as shown in Fig. 10.11. It comprises a reinforced two-layer design to offer greater strength with an average outer surface porosity of 50 μm and inner surface porosity of 20 μm [37].

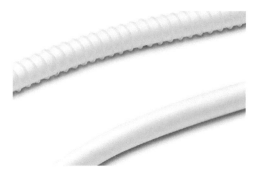

Figure 10.11 Advanta VXT vascular graft [37]. *Source: Image reproduced with permission from Getinge.*

Advanta grafts come in two different configurations [37]:
- Straight
- Tapered

Straight grafts are often used for applications where bypass is needed and where vessel diameter remains constant from the proximal to the distal anastomosis. Tapered grafts are used where flow dynamics need to be altered or when performing bypass between the two different diameter vessels [37].

10.3.2 Vascular patches

The primary intent of vascular patches is to close sutures. In every surgical procedure, closure of sutures needs to be done effectively. Vascular patches are made from synthetic or biological materials and are characterized through excellent biocompatibility.

Getinge offers a range of patches that are made from collagen-coated polyester as shown in Fig. 10.12. All these patches feature reverse "locknit knitting" technique, and the water permeability is less than 5 ml/cm^2/min [38].

10.3.3 Tunneling solutions

Getinge's tunneling solutions consist of the following:
- Vascular Graft Tunneling Instrumentation
- SLIDER GDS

10.3.3.1 Vascular graft tunneling instrumentation

Vascular Graft Tunneling Instrumentation from Getinge is shown in Fig. 10.13 . It is designed to create a subcutaneous tunnel for the placement

Therapeutic applications of medical devices

Figure 10.12 Vascular patches for vascular and cardiothoracic surgery [38]. *Source: Image reproduced with permission from Getinge.*

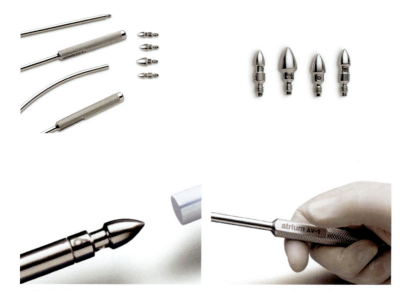

Figure 10.13 Vascular graft tunneling instrumentation [39]. *Source: Image reproduced with permission from Getinge.*

of a vascular graft in both peripheral and AV access applications. While the Tunneling System can be used with a graft that does not feature a Slider Graft Deployment System (GDS), this proprietary Tunneling system is designed to be most effective when used in conjunction with a GDS graft [39].

The tunneling system and its components are made of surgical-grade stainless steel and are designed for multiple reuse. It comes with four tunneling rod configurations to accommodate both AV loop graft placement and bypass graft placement. A thumb tab on each rod provides directional control. It also includes assortment of blunt and pointed tips that feature a large grasping area for easier manipulation. These tips create a tighter tunnel to match graft diameter.

10.3.3.2 SLIDER graft deployment system

SLIDER Graft Deployment System (GDS) from Getinge is shown in Fig. 10.14. It is a patented solution that is used in implanting a vascular graft. It provides fast graft attachment and insertion. The low-profile Vascular Graft Tunneling Instrumentation tip with its thin protection sleeve minimizes trauma, allows for a tight graft tunnel, and minimizes prewetting of the graft. The SLIDER GDS has a double sheath system designed to minimize the risk of contaminating the graft material prior to insertion [40].

Figure 10.14 SLIDER graft deployment system (GDS) [40]. *Source: Image reproduced with permission from Getinge.*

10.3.4 Jetstream atherectomy system

Jetstream atherectomy is a rotational cutting device that is used to treat obstructive peripheral arterial disease. The effectiveness of the device depends on the following variables [41]:
- Type of wire used
- Size of the wire
- Cutter translation speed
- Use of fluoroscopic imaging
- Tactile and auditory senses

As shown in Fig. 10.15, Boston Scientific's Jetstream atherectomy system consists of a single-use catheter with control pod and a reusable, compact console power source that mounts to a standard IV stand. The system is designed to restore flow through the different types of lesion morphologies encountered in peripheral arterial disease as well as treat calcium, plaque, or thrombus. Jetstream atherectomy system is the only atherectomy system with active aspiration that can remove debris, thereby helping in minimizing the risk of distal embolization [42].

10.4 Interventional cardiology

Interventional cardiology is a branch of cardiology that deals specifically with the catheter-based treatment of structural heart diseases. The field includes the diagnosis and treatment of coronary artery disease, vascular disease, and acquired structural heart disease.

10.4.1 WATCHMAN left atrial appendage closure device

WATCHMAN Left Atrial Appendage Closure Device from Boston Scientific shown in Fig. 10.16 is a medical device that can be used to perform left atrial appendage closure as an alternate procedure to OAC therapy that is often used to reduce stroke risk in patients with atrial fibrillation [43].

Left atrial appendage closure is a one-time procedure that reduces the risk of stroke and bleeding in patients with atrial fibrillation patients. WATCHMAN is available in five different sizes and is repositionable and

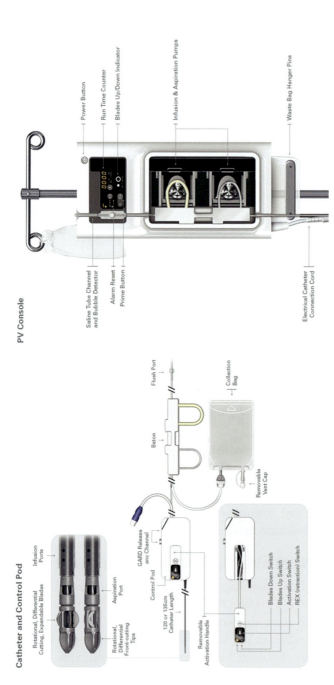

Figure 10.15 Jetstream atherectomy system from Boston scientific [42]. *Source: Image reproduced with permission from Boston Scientific.*

Therapeutic applications of medical devices 187

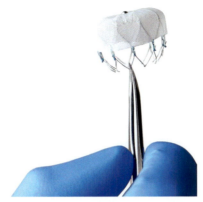

Figure 10.16 WATCHMAN left atrial appendage closure device from Boston scientific [43]. *Source: Image reproduced with permission from Boston Scientific.*

recapturable. The device comes complete with the delivery system and is permanently implanted at distal to the opening of the left atrial appendage. The proximal face minimizes surface area facing the left atrium to reduce postimplant thrombus formation. 160-μm PET membrane promotes the endothelialization process [43].

10.4.2 SYNERGY bioabsorbable polymer stent

SYNERGY bioabsorbable polymer (BP) stent from Boston Scientific shown in Fig. 10.17 is the first FDA-approved drug-eluting stent with abluminal bioabsorbable polymer coating [44]. The SYNERGY BP stent addresses challenges that are associated with permanent polymer stents such as inflammation, neoatherosclerosis, and late stent thrombosis. Once the

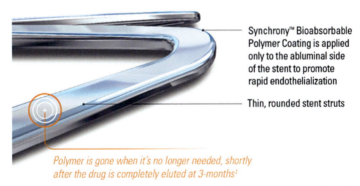

Figure 10.17 SYNERGY bioabsorbable polymer stent from Boston scientific [44]. *Source: Image reproduced with permission from Boston Scientific.*

drug is completely eluted at the end of 3 months, the polymer on the stent is absorbed, thereby providing early healing and freedom from long-term polymer exposure [44].

10.4.3 XIENCE Sierra everolimus eluting coronary stent system

XIENCE Sierra everolimus eluting coronary stent system from Abbott is shown in Fig. 10.18. It is used in situations that require enhanced flexibility, crossability, and pushability. It has the lowest crossing profile and robust radial and longitudinal strength of any drug-eluting stents on the market. It offers postdilatation up to 5.5 mm and retains superior integrity, even at maximum expansion [45].

10.4.4 NC Trek coronary dilation catheter

NC Trek coronary dilation catheter from Abbott shown in Fig. 10.19 is part of the Trek family of balloons [46]. Some of its key features include customized tip, flexible dual tungsten markers, and multilayer crossflex balloon. NC Trek is designed with short steep tapers primarily to focus dilatation on the lesion and not on the surrounding tissue. The design allows minimal shoulder-to-shoulder balloon growth for controlled balloon expansion and outstanding stent apposition [46].

10.4.5 Trek and Mini Trek coronary dilation catheter

Trek and Mini Trek coronary dilation catheters from Abbott shown in Fig. 10.20 are part of Trek family of balloons [47]. Some of the key features include:
- Transitionless tip
- Redesigned hypotube flexible distal shaft

Figure 10.18 XIENCE Sierra everolimus eluting coronary stent system from Abbott [45]. *Source: Image reproduced with permission from Abbott.*

Therapeutic applications of medical devices 189

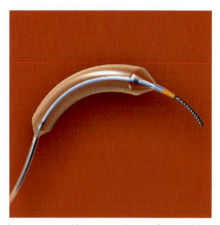

Figure 10.19 NC Trek coronary dilation catheter from Abbott [46]. *Source: Image reproduced with permission from Abbott.*

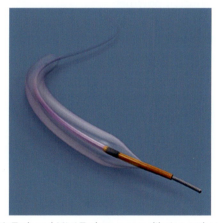

Figure 10.20 Abbott's Trek and Mini Trek coronary dilation catheter [47]. *Source: Image reproduced with permission from Abbott.*

- Multi-layer Crossflex balloon
- Slim seal technology
- Flexible tungsten marker
- Excellent guidewire comfortability
- Ultralow crossing profile
- Outstanding push transmission and lesion access

10.4.6 XIENCE Alpine everolimus eluting coronary stent system

XIENCE Alpine Everolimus Eluting coronary stent system from Abbott shown in Fig. 10.21 is part of the XIENCE family of drug-eluting stents that is used in treating blocked artery. It is designed for complex stent implant procedure known as percutaneous coronary intervention (PCI) [48]. Some of the key features include the following:
- Precision stent placement
- True center tip
- Higher performance catheter
- Durable balloon with flat compliance
- Cobalt chromium in construction
- Flexible stent and delivery system
- Low metal-to-artery ratio
- Coated with fluorinated polymer
- Attracts albumin to surface for thromboresistance
- Minimal inflammation
- Everolimus elution rate matched to restenosis cascade by optimal coating thickness
- Low drug dose
- Broad therapeutic range

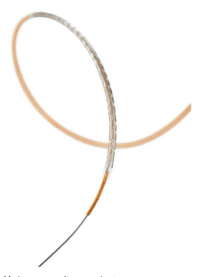

Figure 10.21 XIENCE Alpine everolimus eluting coronary stent system from Abbott [48]. *Source: Image reproduced with permission from Abbott.*

10.5 Thoracic drainage systems

Thoracic drainage systems are designed to remove air, liquids, and solids from the pleural space collected as a result of injury, disease, or surgical procedures.

They are designed in such a way that it creates one-way mechanism only allowing air and fluid out of the pleural space and prevents outside air and fluid from entering into the pleural space. This is accomplished by using an underwater seal. The distal end of the drainage tube is submerged in 2 cm of H_2O. Flexible plastic tubes are inserted through the chest wall and into the pleural space between the fifth and sixth intercostal space in the midaxillary line, venting the space that allows air back-out [49,50].

Teleflex's Pleur-evac chest drainage system are the family of chest drains for thoracic, cardiovascular, trauma, and critical care and uses the most advanced fluid management technology available [51–54].

The Pleur-evac chest drainage system comes in four types:
- Pleur-evac S-1100 Sahara Series (Dry Suction/Dry Seal Control)
- Pleur-evac A-6000 Cactus Series (Dry Suction/Wet Seal Control)
- Pleur-evac A-7000 and A-8000 Rain Series (Wet Suction/Wet Seal Control)
- Pleur-evac Pneumonectomy Unit

Becton Dickinson's PleurX drainage system is a device that drains fluid from abdomen or chest. It helps patient manage recurrent pleural effusions and malignant ascites at home. The system includes an in dwelling catheter and vacuum bottles that allow patients to drain fluid effectively [55].

Getinge has four primary chest drainage system available on the market as shown in Fig. 10.22. They are:
- Express Chest Drain
- Express Mini 500
- Oasis Series
- Ocean Series

Express chest drains offer quick and convenient setup. The Express chest drain has a dry seal one-way valve for patient safety and integrates precision dry suction regulation [56]. In comparison with Express chest

Atrium Express® Dry Seal Chest Drain

Atrium Express Mini 500® Mobile Dry Seal Drain

Atrium Oasis® Dry Suction Water Seal Chest Drain

Atrium Ocean® Wet Suction Water Seal Chest Drain

Figure 10.22 Getinge's chest drainage systems [56–59]. *Source: Images are reproduced with permission from Getinge.*

drains, Express Mini 500 is a small, portable drainage collection unit with a collection volume of 500 mL and an automatic preset suction regulator set at -20 cm H_2O. The Express Mini includes a dry seal one-way valve for patient protection [57].

Oasis drainage series from Getinge comes with a universal water seal technology with air leak monitor for patient air leaking trending. The suction in the system can be adjusted from -10 to -40 cm H_2O. It has internal knock-over nozzles to reduce interchamber fluid spills [58].

Ocean drainage series from Getinge features a graduated water seal that provides convenient assessment of intrathoracic pressure. It has a universal water seal technology with air leak monitoring, an internal knock-over nozzles to minimize interchamber spills, and suction control stopcock for easy adjustment of water bubbling [59].

Smith Medical has comprehensive list of drainage products on market. PORTEX ambulatory chest drainage kit and PORTEX ambulatory chest drainage system are two commercial products available for pneumothorax and hemothorax and drainage of effusions and empyemas [60].

Merit Medical's Aspira drainage system is a tunneled, long-term catheter used for draining accumulated fluid from the pleural or peritoneal cavity to relieve symptoms associated with malignant pleural effusion or malignant ascites [61].

10.6 Prosthetic implants

Implants are medical devices that are placed inside or on the body. Many implants are prosthetics, intended to replace missing body parts. Other implants deliver medications, monitor body functions, or provide support to organs and tissues. Implants can be placed permanently, or they can be removed once they are no longer needed. Some of the key manufacturers of prosthetic implants include the following:
- Stryker Corporation
- Johnson and Johnson
- Otto Bock Healthcare
- Smith and Nephew

10.6.1 Stryker corporation

Stryker Corporation is a leading company that offers a range of implantable orthopedic products. Table 10.4 shows products based on categories [62]:
- Variable angle locking patella plating system
- RIA 2 System
- Femoral Neck System (FNS)
- TRUMATCH® Graft Cage
- ViviGen® & ViviGen Formable Cellular Bone Matrix
- TFN-ADVANCED® Proximal Femoral Nailing System

Table 10.4 Stryker's Implantable Orthopedic Products [62].

Hip Designed for patients who need hip replacement implants for hip arthroplasty	Knee Designed for patients who need total knee replacement	Limb Salvage Designed for patients who need extensive reconstruction of hip joint and/or knee joint
• Accolade II • Anato • Exeter • Anatomic Dual Mobility • Direct Superior Approach • Modular Dual Mobility • Secur-Fit Advanced • Trident • Trident II • X3	• GMRS • Restoris MCK • Triathlon Primary • Triathlon Revision • Triathlon Tritanium • X3	• GMRS

Source: Stryker Corporation

10.6.2 Johnson and Johnson

DePuy Synthes, a subsidiary of Johnson and Johnson (J&J), is a global company that treats orthopedic trauma. Some of J&J commercial products include [63].

10.6.3 Otto Bock

Otto Bock is a global market leader in prosthetics with focuses on finding solutions to knees, feet, socket, and hips. Otto Bock's prosthetic solutions are divided into four areas—(1) prosthetic knee solution, (2) prosthetic foot solution, (3) prosthetic socket solution, and (4) prosthetic hip solutions as shown in Figs. 10.23–10.26, respectively [64]:

10.6.3.1 Prosthetic knee solution

Figure 10.23 Prosthetic knee solutions from otto Bock [64]. *Source: Images are reproduced with permission from Otto Bock.*

10.6.3.2 Prosthetic foot solution

Figure 10.24 Prosthetic foot solutions from otto Bock [64]. *Source: Images are reproduced with permission from Otto Bock.*

10.6.3.3 Prosthetic socket solution

Figure 10.25 Prosthetic socket solutions from otto Bock [64]. *Source: Images are reproduced with permission from Otto Bock.*

10.6.3.4 Prosthetic hip solutions

Helix 3D

Figure 10.26 Prosthetic hip solutions from otto Bock [64]. *Source: Images are reproduced with permission from Otto Bock.*

10.6.4 Smith and Nephew

Smith and Nephew is a world leader in joint replacement systems for knees, hips, and shoulders. The product lines includes knee and hip implant system (Table 10.5), shoulder joints, and ancillary products such as bone cement [65].

Table 10.5 Smith and Nephew's knee and hip implant product line [65].

Knee system	Hip implant system
• Legion/Genesis II total knee system • Journey II active knee solutions	• Anthology hip system • Synergy hip system • SMF femoral hip system • Polarstem femoral hip system • R3 acetabular system • Polarcup Dual mobility hip system • Redapt revision hip system

10.7 In vitro diagnostics

In vitro diagnostics (IVD) are clinical tests that analyze samples taken from the human body. IVD can detect diseases or other conditions and can be used to monitor a person's overall health to help cure, treat, or prevent diseases.

Some of the leading IVD companies include the following:
- Abbott Laboratories
- Bio-Rad Laboratories Inc.
- Danaher Corporation
- Hologic Inc.

10.7.1 Abbott laboratories

M2000 RealTime System from Abbott Laboratories shown in Fig. 10.27 is a highly flexible system that features broad menu of IVD assays [66]. It allows consolidation of PCR testing on a single, reliable, high-performance platform. Some of its key features include the following:
- Flexible sample processing
- Positive ID barcode reader
- Load and go capability
- One platform and multiple uses
- Smart system design
- Custom software

Figure 10.27 Abbott's M2000 RealTime system [66]. *Source: Image reproduced with permission from Abbott.*

10.7.2 Bio-Rad laboratories Inc.

Bio-Rad Laboratories have three types of IVD real-time PCR systems as shown in Fig. 10.28 [67—69]:
- CFX Opus (96 & 384)

Therapeutic applications of medical devices

- CFX Connect
- CFX96 Touch Deep Well

CFX real-time PCR detection systems offer flexibility and capability to deliver sensitive and reliable detection of both single-plex and multiplex real-time PCR reactions. They feature two to five color multiplexing, advanced optical technology, and precise temperature control with thermal gradients [67–69].

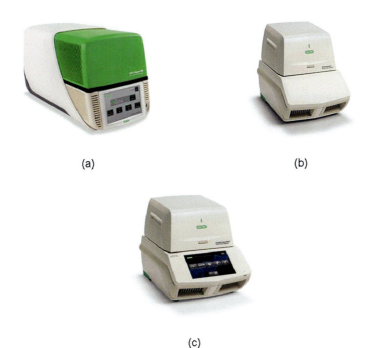

(a) (b)

(c)

Figure 10.28 Bio-rad's real-time PCR detection systems—(1) CFX opus 96, (2) CFX Connect, and (3) CFX96 touch deep well [67–69]. *Source: Images are reproduced with permission from Bio-Rad Laboratories.*

10.7.3 Danaher corporation

Danaher's IVD capabilities are distributed among its three diagnostic businesses—(1) Beckman Coulter, (2) Leica Biosystems, and (3) Radiometer.

Beckman Coulter offers a full range of pre- and postanalytical automation solutions for clinical laboratories. They include the following:

- AutoMate 2500 Family
- Power Link

- Power Express
- DxA 5000 Total Lab Automation

10.7.3.1 Automate 2500 family

AutoMate 2500 Family sample processing system is a high-impact pre- and postanalytical laboratory automation system. Some of its key features include the following [70]:
- Intelligent sample banking (ISB) software to deliver an automation system that fits needs of the laboratory
- Allows streamlining pre- and postanalytical processes to help gain optimal performance and use of resources, eliminating steps between sample receipt and analysis

10.7.3.2 Power link workcell

Beckman Coulter's Power Link Workcell is a scalable workcell that provides support to low-to-mid volume hospital laboratories. It is designed to optimize workflow and expedite delivery of results. Power Link Workcell shares a single specimen between the two analyzers, providing a flexible workflow and eliminating compromises when working with multiple disciplines [71].

10.7.3.3 Power express laboratory automation system

Power Express laboratory automation system provides a total laboratory automation solution to mid- to high-volume laboratories. It is designed to maximize uptime, minimize errors, and optimize workflow. The Power Express lab automation system helps drive sample processing through intelligent sample management allowing for the right tube to be sent at the right time based on testing requirements and instrument workload [72].

10.7.3.4 DxA5000 total laboratory automation system

DxA5000 total laboratory automation system uses a proprietary intelligence that drives rapid and consistent turnaround times for all samples. It performs comprehensive preanalytical sample inspection that helps reduce costs by taking out of the workflow unsuitable samples due to improper specimen collection [73].

10.7.4 Hologic Inc.

Hologic is a leading medical device company that manufactures and supply diagnostic products. Hologic's Panther Scalable Solutions allows expansion of IVD testing menu while adding flexibility, capacity, and walk-away time.

10.7.4.1 Panther system
Panther system as shown in Fig. 10.29 is the foundation of the Hologic's Panther Scalable Solutions. It gives the capability to consolidate menu on a fully automated system, load samples in any order at any time, and decrease turnaround time [74].

10.7.4.2 Panther fusion system
Hologic's Panther system can be upgraded to Panther Fusion system to add more flexibility and capacity. Panther Fusion system as shown in Fig. 10.30 can run real-time PCR, TMA, and RT-TMA assays on a single, fully automated platform. The Panther Fusion increases walkaway time, enhances flexibility, and consolidates testing [75].

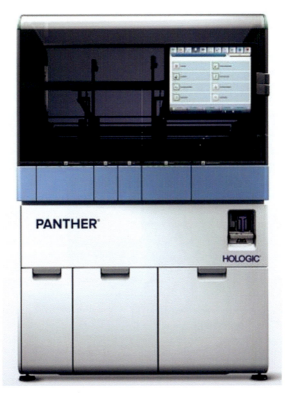

Figure 10.29 Hologic's panther system [74]. *Source: Images are reproduced with permission from Hologic.*

Figure 10.30 Panther fusion system [75]. *Source: Images are reproduced with permission from Hologic.*

10.7.4.3 Panther plus

In high-volume laboratories where there is a need to increase the capacity, Panther Fusion system can be upgraded by adding multitransport unit (MTU) module to Panther Fusion system. The combined system is referred to as Panther Plus as shown in Fig. 10.31 [76].

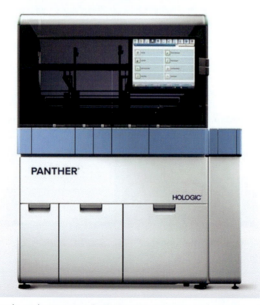

Figure 10.31 Panther plus system [76]. *Source: Images are reproduced with permission from Hologic.*

Figure 10.32 Panther link to connect two panther systems [77]. *Source: Images are reproduced with permission from Hologic.*

10.7.4.4 Panther link
If an existing Panther System needs to connect to other Panther systems as shown in Fig. 10.32 and be able to share inventories, chemicals, and results while monitoring, a link can be established by using Panther Link [77].

References
[1] OVITEX, TELA Bio® <https://www.telabio.com/ovitex.html>.
[2] OVITEX PRS, TELA Bio® <https://www.telabio.com/ovitex.html#ovitex-prs>.
[3] Endoform® Antimicrobial Restorative Bioscaffold, AROA Bio®, <https://aroabio.com/product/endoform/>.
[4] Myriad Matrix™, AROA Bio® <https://aroabio.com/product/myriad-matrix/>.
[5] Myriad Morcells™, AROA Bio®, <https://aroabio.com/product/myriad-morcells/>.
[6] Biosurgery, Tissue Regenix Group, <https://www.tissueregenixus.com/biosurgery/overview?country-selected=true>.
[7] DermaPure, Tissue Regenix Group, <https://www.tissueregenixus.com/biosurgery/dermapure/dermapure-overview/>.
[8] SurgiPure, Tissue Regenix Group, <https://www.tissueregenixus.com/biosurgery/surgipure-xd/>.
[9] A.T. Trott, "Instruments, Suture Materials, and Closure Choices", Wounds and Lacerations, fourth ed., Elsevier, 2012.
[10] Arista Absorbable Hemostat, Becton and Dickinson, <https://www.bd.com/en-us/offerings/capabilities/biosurgery/hemostats/arista-ah-absorbable-hemostat>.
[11] Avitene Microfibrillar Collage Hemostat, Becton and Dickinson, <https://www.bd.com/en-us/offerings/capabilities/biosurgery/hemostats/avitene-microfibrillar-collagen-hemostat>.
[12] Avitene Sheets, Becton and Dickinson, <https://www.bd.com/en-us/offerings/capabilities/biosurgery/hemostats/avitene-sheets>.

[13] Avitene Ultrafoam Collagen Sponge, Becton and Dickinson, <https://www.bd.com/en-us/offerings/capabilities/biosurgery/hemostats/avitene-ultrafoam-collagen-sponge>.
[14] SyringeAvitene Applicators, Becton and Dickinson, <https://www.bd.com/en-us/offerings/capabilities/biosurgery/hemostats/syringeavitene-applicators>.
[15] Progel Pleural Air Leak Sealant, Becton and Dickinson, <https://www.bd.com/en-us/offerings/capabilities/biosurgery/sealants/progel-pleural-air-leak-sealant>.
[16] Tridyne Vascular Sealant, Becton and Dickinson, <https://www.bd.com/en-us/offerings/capabilities/biosurgery/sealants/tridyne-vascular-sealant>.
[17] Vistaseal Fibrin Sealant, Johnson and Johnson, <https://www.jnjmedicaldevices.com/en-US/news-events/ethicon-launches-vistaseal-fibrin-sealant-human-manage-bleeding-during-surgery>.
[18] Cardiac Science Powerheart G3 and G3 Plus, Recently Approved Devices, Food and Drug Administration, 12/07/2018, <https://www.accessdata.fda.gov/cdrh_docs/pdf16/P160033B.pdf>
[19] Cardiac Science Powerheart G3 Pro AED, Recently Approved Devices, Food and Drug Administration, 12/06/2018, <https://www.accessdata.fda.gov/cdrh_docs/pdf16/P160034B.pdf>
[20] Zoll Corporation's Automatic External Defibrillator, Food and Drug Administration, <https://wayback.archive-it.org/7993/20201223181338/https://www.fda.gov/medical-devices/recently-approved-devices/zollr-x-seriesr-r-seriesr-propaqr-md-aed-pror-and-aed-3-blsr-professional-defibrillators-p160022>.
[21] Defibtech Lifeline DDU Automated Defibrillator, Food and Drug Administration, 02/01/2018, <https://wayback.archive-it.org/7993/20201223180939/https://www.fda.gov/medical-devices/recently-approved-devices/lifelinereviver-ecg-and-ddu-automated-defibrillators-p160032>.
[22] Philips HeartStart Onsite Defibrillator, Food and Drug Administration, 06/06/2019, <https://www.fda.gov/medical-devices/recently-approved-devices/heartstart-onsite-defibrillator-model-m5066a-heartstart-home-defibrillator-model-m5068a-primary>.
[23] Philips HeartStart FR3 Defibrillator, Food and Drug Administration, <https://www.fda.gov/medical-devices/recently-approved-devices/philips-heartstart-fr3-defibrillator-primary-battery-rechargeable-battery-charger-rechargeable>.
[24] Philips HeartStart FRx Defibrillator, Food and Drug Administration, <https://www.fda.gov/medical-devices/recently-approved-devices/heartstart-frx-defibrillator-861304-primary-battery-model-m5070a-aviation-frx-battery-989803139301>.
[25] DiamondTemp Ablation Catheter, Medtronic, <https://www.medtronic.com/us-en/healthcare-professionals/products/cardiac-rhythm/ablation-atrial-fibrillation/diamondtemp-ablation-catheters.html>.
[26] Arctic Front Cardiac Cryoablation Catheter, Medtronic, <https://www.medtronic.com/us-en/healthcare-professionals/products/cardiac-rhythm/ablation-atrial-fibrillation/arctic-front-cardiac-cryoablation-catheter.html>.
[27] Pacemakers, Medtronic, <https://www.medtronic.com/us-en/patients/treatments-therapies/pacemakers/our.html>.
[28] Pacemakers, Boston Scientific, <[https://www.bostonscientific.com/en-US/products/pacemakers.html>.
[29] ICD Systems, Medtronic, <https://www.medtronic.com/us-en/healthcare-professionals/products/cardiac-rhythm/icd-systems.html>.
[30] Defibrillators ICD, Boston Scientific, <https://www.bostonscientific.com/en-US/patients/about-your-device/defibrillators-icds.html>.

[31] S-ICD System, Boston Scientific, <https://www.bostonscientific.com/content/dam/lifebeat-online/en/SpecSheets/CRM-389607-AA_EMBLEM_MRI_Patient_Spec_Sheet.pdf>.
[32] ENERGEN ICD, Boston Scientific, <https://www.bostonscientific.com/content/dam/lifebeat-online/en/SpecSheets/CRM-43905-AC_PAE_ENERGEN_ICD_SpecSht_.pdf>.
[33] PUNCTUA ICD, Boston Scientific, <https://www.bostonscientific.com/content/dam/lifebeat-online/en/SpecSheets/CRM-43901-AC_PAE_PUNCTUA_ICD_SpecSht.pdf>.
[34] S. Ravi, E.L. Chaikof, Biomaterials for vascular tissue engineering, Regen. Med. 5 (1) (2010) 107.
[35] Flixene Vascular Graft, Getinge, <https://www.getinge.com/us/product-catalog/flixene-vascular-grafts/>.
[36] Flixene AV Access Vascular Graft, Getinge, <https://www.getinge.com/dam/hospital/documents/english/mcv00088465_reva_flixene_av-access_us-en-us.pdf>.
[37] Advanta VXT Vascular Graft, Getinge, <https://www.getinge.com/us/product-catalog/advanta-vxt-vascular-graft/>.
[38] Patches, Getinge, <https://www.getinge.com/int/products/hospital/vascular-and-cardiothoracic-surgery-solutions/patches/>.
[39] Slider Graft Deployment System, Getinge, <https://www.getinge.com/us/product-catalog/slider-graft-deployment-system/>.
[40] Vascular Graft Tunneling Instrumentation, Getinge, <https://www.getinge.com/us/product-catalog/vascular-graft-tunneler-instrumentation/>.
[41] N.W. Shammas, JETSTREAM atherectomy: a review of technique, tips, and tricks in treating the femoropopliteal lesions, Int. J. Angiol. 24 (2015) 81–86.
[42] Jetstream Atherectomy, Boston Scientific, <https://www.bostonscientific.com/en-US/products/atherectomy-systems/jetstream-atherectomy-system.html>.
[43] Watchman, Boston Scientific, <https://www.bostonscientific.com/en-EU/products/laac-system/watchman.html>.
[44] Bioabsorbable Polymer Stents, Boston Scientific, <https://www.bostonscientific.com/en-US/products/stents–coronary/bioabsorbable-polymer-stent/endotheliazation.html>.
[45] XIENCE Sierra Coronary Stent System, Abbott, <https://www.cardiovascular.abbott/us/en/hcp/products/percutaneous-coronary-intervention/xience-sierra-coronary-stent-system.html>.
[46] NC Trek Coronary Dilation Catheters, Abbott, <https://www.cardiovascular.abbott/us/en/hcp/products/percutaneous-coronary-intervention/nc-trek-coronary-dilation-catheters.html>.
[47] NC Trek Mini Coronary Dilation Catheters, Abbott, <https://www.cardiovascular.abbott/us/en/hcp/products/percutaneous-coronary-intervention/nc-trek-mini-coronary-dilation-catheters.html>.
[48] XIENCE Alpine Drug Eluting Stent, Abbott, <https://www.cardiovascular.abbott/us/en/hcp/products/percutaneous-coronary-intervention/xience-alpine-drug-eluting-stent.html>.
[49] E.R. Munnell, Thoracic drainage, Ann. Thorac. Surg. 63 (1997) 1497–1502.
[50] S. McDermott, D.A. Levis, R.A. Arellano, Chest drainage, Semin. Intervent. Radiol. 29 (2012) 247–255.
[51] Dy Suction Dry Seal, Chest Drain, Teleflex, <https://www.teleflex.com/usa/en/product-areas/surgical/cardiovascular/chest-drainage/dry-suction-dry-seal/>.

[52] Dry Suction Wet Seal, Chest Drain, Teleflex, <https://www.teleflex.com/usa/en/product-areas/surgical/cardiovascular/chest-drainage/dry-suction-wet-seal/>.
[53] Wet Suction Wet Seal, Chest Drain, Teleflex, <https://www.teleflex.com/usa/en/product-areas/surgical/cardiovascular/chest-drainage/wet-suction-wet-seal/>.
[54] Pneumonectomy, Chest Drain, Teleflex, <https://www.teleflex.com/usa/en/product-areas/surgical/cardiovascular/chest-drainage/pneumonectomy/>.
[55] Pleurx Drainage System, Becton Dickinson, <https://www.bd.com/en-us/offerings/capabilities/drainage/peritoneal-and-pleural-drainage/about-the-pleurx-drainage-system/pleurx-drainage-system>.
[56] Express Chest Drain, Getinge, <https://www.getinge.com/us/product-catalog/express-dry-seal-chest-drain/>.
[57] Express Mini 500 Chest Drain, Getinge, <https://www.getinge.com/us/product-catalog/express-mini-500-mobile-dry-seal-drain/>.
[58] Oasis Series Chest Drain, Getinge, <https://www.getinge.com/us/product-catalog/oasis-dry-suction-water-seal-chest-drain/>.
[59] Ocean Series Chest Drain, Getinge, <https://www.getinge.com/us/product-catalog/ocean-wet-suction-water-seal-chest-drain/>.
[60] Drainage Systems, Smiths Medical, <https://www.smiths-medical.com/en-us/products/drainage-systems>.
[61] Aspira Drainage System, Merit Medical Systems, <https://www.merit.com/peripheral-intervention/drainage/complete-solutions/aspira-drainage-system/>.
[62] Joint Replacement, Stryker, <https://www.stryker.com/us/en/portfolios/orthopaedics/joint-replacement.html>.
[63] Prosthetics, Trauma, Johnson and Johnson, <https://www.jnjmedicaldevices.com/en-US/specialty/trauma>.
[64] Lower Limb Prosthetics, Otto Bock, <https://www.ottobockus.com/prosthetics/lower-limb-prosthetics/solution-overview/>.
[65] Orthopedic Reconstruction, Smith and Nephew, <https://www.smith-nephew.com/about-us/what-we-do/orthopaedic-reconstruction/>.
[66] M2000 Real-Time System, Abbott, <https://www.molecular.abbott/us/en/products/instrumentation/m2000-realtime-system>.
[67] CFX Opus Real-Time PCR Systems, Bio-Rad Laboratories, <https://www.bio-rad.com/en-us/product/cfx-opus-real-time-pcr-systems?ID=QBJBMKRT8IG9>.
[68] CFX Connect Real-Time PCR Detection System, Bio-Rad Laboratories, <https://www.bio-rad.com/en-us/product/cfx-connect-real-time-pcr-detection-system?ID=LN5TFG15>.
[69] CFX96 Real-Time PCR Detection System, Bio-Rad Laboratories, <https://www.bio-rad.com/en-us/product/cfx96-touch-deep-well-real-time-pcr-detection-system?ID=LZJTUJ15>.
[70] Automate 2500 Family Sample Processing System, Beckman Coulter, <https://www.beckmancoulter.com/en/products/automation/automate-2500-family-sample-processing-systems>.
[71] Power Link, Beckman Coulter, <https://www.beckmancoulter.com/en/products/automation/power-link>.
[72] Power Express Laboratory Automation System, Beckman Coulter, <https://www.beckmancoulter.com/en/products/automation/power-express-laboratory-automation-system>.
[73] DXA 5000 Laboratory Automation System, Beckman Coulter, <https://www.beckmancoulter.com/en/products/automation/dxa-5000-lab-automation-system#/overview>.

[74] Panther System, Hologic Inc., <https://www.hologic.com/hologic-products/diagnostic-solutions/panther-scalable-solutions/panther-system>.
[75] Panther Fusion System, Hologic Inc., <https://www.hologic.com/hologic-products/diagnostic-solutions/Panther-Scalable-Solutions/panther-fusion-system>.
[76] Panther Plus, Hologic Inc., <https://www.hologic.com/hologic-products/diagnostic-solutions/Panther-Scalable-Solutions/panther-plus>.
[77] Panther Link, Hologic Inc., <https://www.hologic.com/hologic-products/diagnostic-solutions/Panther-Scalable-Solutions/panther-link>.

CHAPTER ELEVEN

Applications of biobased polymers in medical devices

Contents

11.1 Introduction	209
11.2 Implantable medical devices	210
11.2.1 EVONIK's biodegradable polymers	210
11.2.1.1 RESOMER custom biodegradable polymers	210
11.2.1.2 VESTAKEEP polyether ether ketone	211
11.2.2 REGENESORB	213
11.2.2.1 MICRORAPTOR REGENESORB	214
11.2.3 Biocomposite	214
11.2.4 Biocryl Rapide biocomposite	215
11.2.5 GENESYS biocomposite	215
11.2.6 DuoSorb biocomposite	215
11.2.7 Bilok biocomposite	216
11.2.8 Biosteon biocomposite	216
11.2.9 Solviva Biomaterials	216
11.2.9.1 Eviva polysulfone	216
11.2.9.2 Veriva polyphenylsulfone	216
11.2.9.3 Zeniva polyether ether ketone	217
11.2.10 Fused filament fabrication 3D-printed polyether ether ketone implant	217
11.2.11 VICRYL suture	217
11.2.12 TephaFlex absorbable monofilament sutures	218
11.3 Diagnostics and labware devices	218
11.3.1 ALTUGLAS Rnew	218
11.4 Healthcare electronic devices	218
11.4.1 Solvay Kalix 2000 series	218
11.5 Medical mask and medical tubing	219
11.5.1 ARKEMA Rilsan polyamide 11 resin	219
References	219

11.1 Introduction

There has been a significant leap in the development of biodegradable and biobased polymers. Biodegradable polymers such as polylactic acid (PLA), polycaprolactone (PCL), polylactic-co-glycolic acid (PLGA), and

polyhydroxyalkanoates (PHA) as their copolymers are being frequently used in medical devices because of their excellent biocompatibility, compatibility to different sterilization methods, outstanding toughness and stiffness, resistance to chemicals, and ease of processability. Resin manufacturers Arkema and Solvay are able to convert fossil-based polymers such as polyether ether ketone (PEEK), polysulfone (PSU), and polyphenylsulfone (PPSU) into more renewable source. This chapter will review applications of biobased polymers in medical devices focusing on four different areas:
1. Implantable medical devices
2. Diagnostic and labware
3. Healthcare electronic devices
4. Medical mask and medical tubing

11.2 Implantable medical devices
11.2.1 EVONIK's biodegradable polymers
11.2.1.1 RESOMER custom biodegradable polymers

RESOMER refers to portfolio of standard and custom biodegradable polymers developed by EVONIK. They include lactide, lactide-PEG, caprolactone, dioxanone, glycolide, and composite-based biodegradable polymers [1]. RESOMER can either be semicrystalline or amorphous with superior mechanical and chemical properties making them an ideal candidate for medical device applications such as orthopedics, spine, dental, cardiovascular, wound healing, suture anchors, and interference screws as shown in Fig. 11.1 [2]. The degradation ranges from 1 month to more than 4 years.

RESOMER composite-based biodegradable polymers offer mechanical properties that match the natural bone and minimize stress shielding for

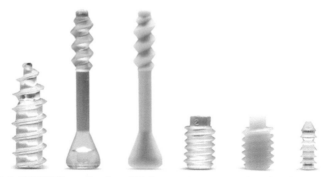

Figure 11.1 RESOMER applications [2]. *Source: Image reproduced with permission from EVONIK.*

faster bone healing with reduced inflammation. When blended with calcium phosphate—based ceramic additives, they offer improved osteoconductive properties. The material is supplied as ready-to-use pellets that are easy to process. Similarly, when a RESOMER polymer is combined with lactide—PEG copolymer, it produces a triblock copolymer, which has mechanical properties of RESOMER polymer and hydrophobic and hydrophilic properties from lactide and PEG, respectively. RESOMER polymers can degrade six times faster [1—3].

11.2.1.2 VESTAKEEP polyether ether ketone
EVONIK developed VESTAKEEP to address the need of medical device industry for resins that can be used as a replacement to metal in medical implant applications. VESTAKEEP has a high fatigue resistance and toughness, which makes it a right candidate for applications in orthopedic, spine, sports medicine, cardiovascular, extremities, craniomaxillofacial, dental, and oncology business areas [4].

Some of the other advantages of VESTAKEEP include the following:
- Biocompatibility
- Biostability
- Sterilization compatibility
- Resistant to chemicals
- Modulus similar to bone
- Metal-free
- Light weight and low thermal conductivity
- Injection molding and extrusion compatible
- Low water absorption
- Easy to machine
- Good processability
- Lower stress-shielding effect

EVONIK offers three grades that cover a range of medical applications [5]:
1. VESTAKEEP i-grades
2. VESTAKEEP care grades
3. VESTAKEEP dental grades

VESTAKEEP i-grades
EVONIK developed VESTAKEEP i-grade as a solution for permanent implants. Its superior biocompatibility and excellent mechanical properties combined with high purity as a result of Evonik's quality control makes

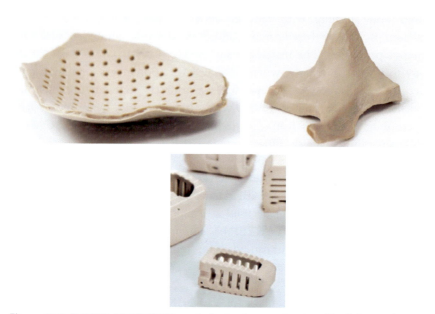

Figure 11.2 EVONIK's VESTAKEEP i-grade for implant applications [6–8]. *Source: Images are reproduced with permission from EVONIK.*

VESTAKEEP i-grade an ideal material for long-term human implants. Some of the application areas include spinal cages, stents, heart valves, facial implants for facial bone fractures, access ports, suture anchors, interference screws, and small joints as shown in Fig. 11.2 [6–8].

VESTAKEEP care grades

VESTAKEEP care grades are EVONIK's solution for medical device applications that include metal replacement, parts for housing and surgical instruments, gear wheels and other functional units, and durable medical equipment. They have excellent biocompatibility, resistance to all commonly used sterilization methods, X-ray transparency, excellent fatigue properties, mechanical properties close to cortical bone, high chemical resistance, very good resistance to wear, good electrical insulation properties, excellent biostability, and high-dimensional stability [5].

VESTAKEEP dental grades

VESTAKEEP dental grades provide metal-free solution for outstanding wear comfort. This grade is used for crown and bridges, cervical gingiva formers, temporary and permanent abutments, attachments restorations, partial dentures and transversal connectors, occlusal splints, inlay bridges, telescopic crowns, dentures, and healing caps as shown in Fig. 11.3 [9].

Applications of biobased polymers in medical devices 213

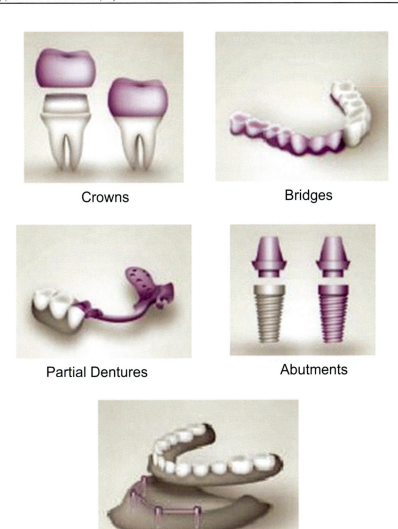

Crowns

Bridges

Partial Dentures

Abutments

Bar restorations

Figure 11.3 EVONIK's VESTAKEEP dental grade applications [9]. *Source: Images reproduced with permission from EVONIK*

11.2.2 REGENESORB

The REGENESORB material from Smith and Nephew is a new bioabsorbable biocomposite material comprised of the copolymer PLGA (poly (L-lactide co-glycolide)) combined with two fillers, calcium sulfate and betatricalcium phosphate (β-TCP) in a ratio of 65:20:15, respectively. Both calcium-based fillers have been individually demonstrated to be osteoconductive [10–12].

11.2.2.1 MICRORAPTOR REGENESORB

MICRORAPTOR from Smith and Nephew shown in Fig. 11.4 is a micro-class anchor with a shallow drill depth that can be completely absorbed and replaced by bone in 24 months while providing a solid finished construct [13].

11.2.3 Biocomposite

Arthrex has developed a new absorbable composite interference screw for graft fixation in anterior cruciate ligament (ACL) and posterior cruciate ligament (PCL) reconstruction procedures [14,15]. The biocomposite interference screw is a combination of 70% poly(L-lactide-co-D, L-lactide) (PLDLA) and 30% biphasic calcium phosphate (BCP) and is intended for use as a fixation device for bone-patellar tendon-bone (BTB) and soft tissue grafts during ACL and PCL reconstruction procedures [16].

PLDLA properties include the following:
- High absorbing predictability over time
- Greatly reduces chance of osteolytic lesions frequently seen with rapidly absorbing PGA polymers and copolymers
- No crystalline degradation product build-up at the implant site
- Safe biodegradable polymer and widely studied

The interference screw is a direct tendon-to-bone interference fixation device. It is a typical compression fixation device, which relies on the screw threads to engage and compress the graft for fixation. This device has been widely used with multiple-looped hamstring tendon grafts in cruciate ligament reconstruction.

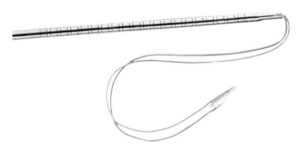

Figure 11.4 Smith and Nephew's MICRORAPTOR REGENESORB [13]. *Source: Images are reproduced with permission from Smith and Nephew.*

11.2.4 Biocryl Rapide biocomposite

BIOCRYL RAPIDE from Johnson and Johnson is a biocomposite material that uses a proprietary microparticle dispersion (MPD) manufacturing process providing a homogenous dispersion of the composite materials, while ensuring β-TCP is in close apposition with the surrounding bone tissues [16].

The MILAGRO ADVANCE interference screw is made from BIOCRYL RAPIDE biocomposite material [17].

11.2.5 GENESYS biocomposite

GENESYS Biocomposite material from CONMED Linvatec is produced using a two-step process. In the first step, the GENESYS material undergoes a patented microfiltration process to get the smallest-sized particles, β-TCP, also referred to as microTCP. These small-sized particles are then uniformly embedded through the 96L/4D PLA bioabsorbable polymer producing a biocomposite material. This biocomposite material is then used to manufacture GENESYS Matryx interference screw [18].

GENESYS Matryx interference screw is used for fixation of ACL and PCL grafts. Screws can deliver strong initial fixation during the critical healing period and provide a scaffold to enable bone-in-growth during subsequent resorption period. It has a low profile rounded head, which helps in reducing trauma to graft and improves femoral aperture fixation. Threads on the tip extend to the distal tip, easing screw insertion [18].

11.2.6 DuoSorb biocomposite

DuoSorb biocomposite material from Zimmer Biomet is a blend of 30% β-TCP and 70% PLA. β-TCP in the mix is designed to provide a scaffold for bone ingrowth. The ComposiTCP is Zimmer Biomet's first biocomposite, absorbable suture anchor system shown in Figure 11.5 is used for securing soft tissue to bone [19]. These anchors are osteoconductive and provide strong fixation.

Figure 11.5 Zimmer biomet ComposiTCP suture anchor system [19]. *Source: Image reproduced with permission from Zimmer Biomet.*

11.2.7 Bilok biocomposite

Bilok Bbiocomposite from Biocomposites Inc. is a blend of β- β-TCP and poly(L-lactic acid) (PLLA) produced using a proprietary process, which maximizes component integrity. Bilok interference screws are made from Bilok biocomposite and used in the reconstruction of ACL and PCL and suture anchors for rotator cuff repairs [20].

11.2.8 Biosteon biocomposite

Bisteon Bbiocomposite from Stryker is a blend of calcium hydroxyapatite and PLLA and is used to produce Biosteon IntraLine suture anchors using proprietary molding process. IntraLine suture anchors are intended for fixation of soft tissue to bone in orthopedic procedures [21,22].
Biosteon features includes [22]:
- Wedge-shaped design provides easier insertion and excellent fixation.
- Rounded threaded design provides protection to the graft during insertion.
- Cruciate driver design allows for an even distribution of force.
- Reduction in tunnel widening improves implant to bone integration.

11.2.9 Solviva Biomaterials

Solviva Biomaterials are family of specialty polymers from Solvay that target industry demand for metal alternatives in areas including orthopedic, cardiovascular, spine, sports medicine, and trauma [23]. The list of biomaterials includes the following:
- Eviva Polysulfone (PSU)
- Veriva Polyphenylsulfone (PPSU)
- Zeniva Polyether Ether Ketone (PEEK)

11.2.9.1 Eviva polysulfone
Eviva PSU is a transparent polymer that offers toughness and strong biocompatibility. It is dimensionally stable and does not require machining. It is compatible with all forms of sterilization and does not show any evidence of cytotoxicity, sensitization, and intracutaneous reactivity. It does not absorb fluids and is artifact-free for effective MRI imaging [23,24].

11.2.9.2 Veriva polyphenylsulfone
Veriva PPSU is a transparent polymer that offers toughness and excellent biocompatibility. Some of the other properties include the following [23]:

- High flexural modulus, impact resistance, and durability
- Resistance to chemicals and harsh disinfectants
- Heat resistance
- Compatible with all forms of sterilization

11.2.9.3 Zeniva polyether ether ketone

Zeniva PEEK is a specialty polymer from SOLVIVA line of biomaterials, which offers high dimensional stability, strength, and stiffness. Some of the other properties include the following [23]:
- Modulus of elasticity similar to that of cortical bone
- Excellent biocompatibility
- Excellent fatigue resistance
- Eliminates risk of allergic reactions to heavy metals

Okani Medical Technology has developed durable all-polymer knee implant from Solvay's Zeniva PEEK. Other applications include spinal and orthopedic implants [25–27].

11.2.10 Fused filament fabrication 3D-printed polyether ether ketone implant

FossiLabs, a subsidiary of Curiteva Inc., developed fused filament fabrication (FFF) 3D-printed bone-like scaffolding structures using a porous PEEK material. This approach would benefit existing implantable devices where bone growth may be desired, providing faster osseointegration.

This approach is intended primarily for use with spacers and cages for the spine [28].

Following is a two-stage process:
- The first stage is to identify solid and controlled bone-like macroporosity regions
- The second stage is to identify desired bone growth areas can then and 3D printed in PEEK

11.2.11 VICRYL suture

VICRYL Suture from Johnson and Johnson is a synthetic and absorbable suture coated with lactide and glycolide copolymer plus calcium stearate. It passes through tissue readily with minimal drag and facilitates ease of handling and smooth tie-down [29].

11.2.12 TephaFlex absorbable monofilament sutures

Absorbable Monofilament Sutures from Tepha, a division of Becton Dickinson are produced using TephaFLEX, a proprietary PHA material offering outstanding tensile strength and prolonged strength retention with improved flexibility. Sutures are designed to be smooth allowing them to slide through tissues easily [30].

11.3 Diagnostics and labware devices
11.3.1 ALTUGLAS Rnew

Altuglas Rnew is a patent pending blend of polymethyl methacrylate and PLA with biobased content $\geq 25\%$, polymers 100% produced from plant sugar.

It is transparent and offers outstanding impact performance and chemical resistance. Some of the other properties include the following [31–33]:
- Can be gamma-sterilized and is also stable to other sterilization techniques
- Improved melt processability
- Increased toughness and impact resistance
- Excellent alternative for chemical resistance to alcohol, oils, lipids, cleaners, and antiseptics
- Light transmittance is 92%

Altuglas Rnew is used in applications such as diagnostics, lab ware, fluid suction, reservoir and fluid collection apparatus, and other general care devices [34].

11.4 Healthcare electronic devices
11.4.1 Solvay Kalix 2000 series

Solvay's Kalix 2000 series are biobased polymers containing monomers that come from the sebacic acid chain, which is derived from nonfood competing and GMO-free castor oil [35].

Kalix2000 series consists of three product series as shown in Table 11.1 [35].

Table 11.1 Kalix 2000 series [35].

Product series	Bio-based content
Kalix 2000 base series	61%
Kalix 2955	27%
Kalix 2855	27%

Kalix 2000 series (Kalix 2955 and Kalix 2855) are two biosourced polyamides with glass fiber reinforcement ranging between 50% and 55% by weight. This material is used for applications where high strength and stiffness, good impact resistance, and excellent dimensional stability after molding are needed. Its low viscosity and excellent flow properties make the material ideal for filling parts with thin-walled sections such as those encountered in the mobile electronics industry [36].

11.5 Medical mask and medical tubing

11.5.1 ARKEMA Rilsan polyamide 11 resin

Arkema's Rilsan polyamide 11 resins are high-performance, biobased transparent polyamides derived from the castor plant [37]. Below is the complete production life cycle steps [38]:
- Step 1: Harvesting castor plant to get castor seeds
- Step 2: Grinding castor seeds to produce castor oil
- Step 3: Monomer synthesis of castor oil to get Amino 11
- Step 4: Polymerization of Amino 11

Arkema's Rilsan Clear line of polymers comes in two different grades depending on their percent biobased content [37]:
- Rilsan Clear G850 Rnew (45% biobased content)
- Rilsan Clear G820 Rnew (65% biobased content)

Rilsan Clear polymers offer an outstanding transparency (91% light transmittance), lightweight, flexibility, high temperature resistance, good surface and wear resistance, good chemical resistance, BPA, and plasticizer free and sterilizable. They are typically used in BPA-free respiratory masks and medical tubing [38].

Rilsan MED polyamide 11 offers improved elastic modulus and flexibility, biocompatibility, great chemical resistance, excellent resistance to sterilization methods, and superior dimensional stability. They are 100% biobased advanced materials, derived from the castor plant, and used in applications that require the strength and performance characteristics of a true thermoplastic, while also requiring sufficient flexibility and elongation [39].

References

[1] EVONIK Resomer Biodegradable Polymer. https://healthcare.evonik.com/en/medical-devices/biodegradable-materials/resomer-portfolio.
[2] EVONIK Resomer Bioresorbable Polymers for Medical Devices. https://healthcare.evonik.com/en/medical-devices/biodegradable-materials/resomer-portfolio/standard-polymers.

[3] EVONIK Resomer Composite Bone Matching Biomaterial Brochure. https://healthcare.evonik.com/product/health-care/downloads/evonik-resomer-composite-bone-matching-biomaterials-brochure.pdf.
[4] VESTAKEEP PEEK Polymer now Used in more than 80 Medical Devices, Plastics Today, March 2017. https://www.plasticstoday.com/medical/vestakeep-peek-polymer-now-used-more-80-medical-devices.
[5] VESTAKEEP PEEK. EVONIK. https://medical.evonik.com/en/high-performance-polymers/vestakeep-peek.
[6] PEEK Biomaterial Implants. EVONIK. https://medical.evonik.com/en/peek-biomaterial-implants.
[7] EVONIK Media Download. https://medical.evonik.com/en/media/downloads.
[8] PEEK Biomaterial Implant. Media Download. EVONIK. https://medical.evonik.com/product/peek-industrial/Downloads/vestakeep-peek-biomaterials-for-medical-implants.pdf.
[9] VESTAKEEP PEEK Biomaterial. Modern Plastics. http://www.modernplastics.com/wp-content/uploads/2015/05/VESTAKEEP-PEEK-Biomaterials-08-2015.pdf.
[10] Regenesorb. Smith and Nephew. https://www.smith-nephew.com/puerto-rico-en/products/all-products/regenesorb/.
[11] D. Hak, The use of osteoconductive bone graft substitutes in orthopaedic trauma, J. Am. Acad. Orthop. Surg. 15 (2007) 525—536.
[12] D.C. Allison, A.W. Lindberg, B. Samimi, R. Mirzayan, L.R. Menendez, A comparison of mineral bone graft substitutes for bone defects, US Oncol. Hematol. 7 (1) (2011) 38.
[13] Microraptors Regenesorb. Smith and Nephew. https://www.smith-nephew.com/professional/products/all-products/microraptorsntm-regenesorbsntm/.
[14] Interference Screw. Anthrex. https://www.arthrex.com/knee/interference-screws/related-science.
[15] Biocomposite Interference Screws. Anthrex. https://www.arthrex.com/knee/fastthread-biocomposite-interference-screws.
[16] Biocomposite Interference Screw a Stronger Turn in ACL PCL Reconstruction. Anthrex. https://www.arthrex.com/resources/brochures/sjjfPvkEEeCRTQBQVo-RHOw/biocomposite-interference-screw-a-stronger-turn-in-acl-pcl-reconstruction.
[17] Depuy Synthes. Medical Expo. https://www.medicalexpo.com/prod/depuy-synthes/product-79814-498676.html.
[18] Genesys™ Matryx™ Interference Screw. https://www.conmed.com/en/medical-specialties/orthopedics/knee/ligament-reconstruction/mpfl/fixation/genesys-matryx-interference-screw.
[19] ComposiTCP Suture Anchors. Zimmer Biomet. https://www.zimmerbiomet.com/medical-professionals/sports-medicine/product/compositcp-suture-anchors.html.
[20] Bilok. Biocomposites Inc. https://www.biocomposites.com/our-products/bilok/.
[21] Biosteon. Stryker. https://www.strykermeded.com/medical-devices/sports-medicine/implants/biosteon/.
[22] Biosteon. Biocomposites Inc. https://www.biocomposites.com/our-products/biosteon/.
[23] Implantable Devices. Solvay. https://www.solvay.com/en/chemical-categories/specialty-polymers/healthcare/implantable-devices.
[24] Solvay: Recent Medical Applications of Sulfone Polymer Technology, Medical Plastics News, July 2013. https://www.medicalplasticsnews.com/news/medical-plastics-technology-news/polysulfones-in/.
[25] Solvay Zeniva PEEK Enables Durable All Polymer Knee Implant Okani Medical Technology. https://www.solvay.com/en/press-release/solvays-zeniva-peek-enables-durable-all-polymer-knee-implant-okani-medical-technology.

[26] Solvay Develops New PEEK Grades for Medical Implantable Products, Plastics Today, March 2017. https://www.plasticstoday.com/medical/solvay-develops-new-peek-grades-medical-implantable-products.
[27] Use of Plastics in Medical Implants are Soaring, Plastics Today, October 2012. https://www.plasticstoday.com/use-plastics-medical-implants-soaring.
[28] A. Essop, Fossilabs Offers Enhanced PEEK Medical Devices with 3D Printed Porous Bone-like Structures, January 2020. https://3dprintingindustry.com/news/fossilabs-offers-enhanced-peek-medical-devices-with-3d-printed-porous-bone-like-structures-166795/>.
[29] Vicryl Sutures. Johnson and Johnson. https://www.jnjmedicaldevices.com/en-US/product/coated-vicryl-polyglactin-910-suture.
[30] TephaFlex Monofilament Sutures. Tepha Medical. https://www.tepha.com/products/monofilament-suture/.
[31] Biobased Plexiglas Aimed at Disposable Medical Device Market, Plastics Today, June 2011. https://www.plasticstoday.com/bio-based-plexiglas-aimed-disposable-medical-devices-market.
[32] Plexiglas Rnew Biobased Acrylic Resins. https://www.plexiglas.com/en/media/news/news/Plexiglas-Rnew-Bio-based-Acrylic-Resins-from-Arkema/.
[33] New Biobased Alloy Plexiglas Rnew Resin with NatureWorks Ingeo Biopolymer Offers Exceptional Performance Characteristics, CECA Chemicals, Arkema, March 2012. https://www.cecachemicals.com/en/media/news/New-Bio-based-Alloy-Plexiglas-Rnew-Resin-with-NatureWorks-Ingeo-biopolymer-offers-exceptional-performance-characteristics-00001/.
[34] PMMA Medical — ALTUGLAS acrylic resins. ALTUGLAS. https://www.altuglas.com/en/resins/resins-by-performance/PMMA-Medical-Altuglas-acrylic-resins/.
[35] Kalix HPPA. Solvay. https://www.solvay.com/en/brands/kalix-hppa.
[36] Kalix 2955. Solvay. https://www.solvay.com/en/product/kalix-2955.
[37] Rilsan Polyamide 11 Resin. Arkema. https://www.arkema.com/global/en/products/product-finder/product/technicalpolymers/rilsan-family-products/rilsan-pa11/.
[38] Rilsan Clear — Transparent Polyamides. Arkema. https://www.extrememateriels-arkema.com/en/product-families/rilsan-polyamide-11-family/rilsan-clear.
[39] Rilsan MED. Product Literature. Arkema. https://www.extrememateriels-arkema.com/en/markets-and-applications/consumer-goods-and-healthcare/healthcare-solutions-overview/products/#RILSAN.

CHAPTER TWELVE

Regulatory requirements for medical devices

Contents

12.1 Overview	223
12.2 The Food and Drug Administration	224
12.2.1 Who is the Food and Drug Administration	224
12.2.1.1 Medical device regulations	*224*
12.2.1.2 Establishment registration and medical device listing	*225*
12.2.1.3 Premarket notification 510(k)	*226*
12.2.1.4 Premarket approval	*226*
12.2.1.5 Investigational device exemption	*228*
12.2.1.6 Quality system regulation	*228*
12.2.1.7 Labeling requirement	*231*
12.2.1.8 Medical Device Reporting	*232*
12.3 European Union Commission	233
12.3.1 Medical device regulations	234
12.4 Other global agencies	234
12.4.1 European Medicines Agency	234
12.4.2 National Medical Products Administration, China	235
12.4.3 Central Drugs and Standard Control Organization, India	237
12.4.4 Health Canada	238
12.5 International Medical Device Regulators Forum	238
References	239

12.1 Overview

Every medical device manufacturer is required to comply with regulations of the country devices that are intended for use. The main goal of regulatory compliance is risk prevention. This allows manufacturers to identify, mitigate, and eliminate risks during product lifecycle and ensures medical devices are safe for patients. Depending on the country, regulatory approval process may vary. In the United States, medical devices are regulated by the Food and Drug Administration (FDA) with an aim to ensure safety and effectiveness of the devices. This chapter will review roles of the FDA and other global agencies that regulate medical devices.

Applications of Polymers and Plastics in Medical Devices
ISBN: 978-0-12-820980-6
https://doi.org/10.1016/B978-0-12-820980-6.00001-1

© 2022 Elsevier Inc.
All rights reserved.

12.2 The Food and Drug Administration

12.2.1 Who is the Food and Drug Administration

The FDA is a federal agency in the United States that regulates products critical to health and safety of the public.

Some of its roles and responsibilities include the following [1—4]:
- Regulates development of human and veterinary biological products
- Regulates development of medical devices
- Regulates cosmetics and devices that emit radiation
- Regulates manufacturing, marketing, and distribution of tobacco products
- Participates in the development and use of standards

12.2.1.1 Medical device regulations

The US FDA regulates all medical devices marketed in the United States. Medical devices are classified based on their risks and regulatory controls required for adequate safety and performance. According to the FDA, medical devices are classified into three categories based on the degree of risk they pose to the patient and/or user [5—7]:
- Class I
- Class II
- Class III

Congress passed Federal Food Drug & Cosmetic Act (FFDCA) providing the FDA an authority to regulate both medical devices and radiation-emitting products. The FDA develops, publishes, and implements regulations that apply to medical devices and radiation-emitting products under FFDCA [8].

Rules, proposed rules, notices of federal agencies and organizations, and presidential executive orders and documents are published in official publication known as Federal Register [8]. All proposed rules are initially subjected to public comments before they are finalized and entered into the printed version annually. This printed version is referred to as Code of Federal Regulations (CFR).

CFR is divided into 50 titles that represent areas that are subjected to regulations. Medical device and radiation-emitting product regulations are listed in Title 21 of CFR Parts 800-1299. Overview of Title 21 CFR Parts and their corresponding areas is listed in Table 12.1 [8].

Table 12.1 Overview of title 21 CFR parts and their corresponding areas [8].

Parts	Area
1–99	• Product jurisdictions
	• Protection of human subjects
	• Institutional review boards
100–799	• Food
	• Human and animal drugs
	• Biologics
	• Cosmetics
800–1299	• Medical devices and radiation-emitting products
1300–1499	• Controlled substances

Source: Food & Drug Administration, Republished on 04-Jan-2022.

12.2.1.2 Establishment registration and medical device listing

Manufacturers and distributors, whether they are domestic or foreign involved in production and distribution of medical devices intended for use in the United States are required to register their establishments annually with the FDA. This process is referred to as establishment registration and is governed by Code of Federal Regulations, Title 21 CFR Part 807 [9].

In addition to registering their establishments with the FDA, they are also required to list devices that are made there and the activities that are being performed on those devices. Additional information such as FDA premarket submission number is also needed at the time of registration if a device requires a premarket submission before being marketed in the United States [9].

Each establishment and distributor are required to provide the following information for each device:
- Manufacturers
- Contract manufacturers
- Contract sterilizers
- Repackagers and relabelers
- Specification developers
- Reprocessors single-use devices
- Remanufacturer
- Manufacturers of accessories and components sold directly to the end user
- US manufacturers of "export only" devices

12.2.1.3 Premarket notification 510(k)

Premarket notification 510(k) submission is required for manufacturers and distributors of medical devices that do not require premarket approval (PMA) and are not exempted from 510(k) requirements. This submission is made to FDA to demonstrate that the device is safe and effective and equivalent the predicate device [10].

Premarket notification 510(k) process requires manufacturers to provide substantial equivalence to another legally US marketed device. By substantial equivalence, the FDA requires the manufacturer to show that the new device is safe, effective, and equivalent to the predicate device. A device is substantially equivalent if the intended use is the same as the predicate and the technological characteristics are the same as the predicate. It is also substantially equivalence if the intended use is the same as the predicate but has different technological characteristics and does not raise different questions of safety and effectiveness [10].

According to the FDA, four types of manufacturers require 510(k) submission [10]:
1. Domestic manufacturers that are introducing a device to the US market
2. Specification developers introducing a device to the US market
3. Repackages or relabelers who make significant change to the labeling
4. Foreign manufacturers or exporters introducing device to the US market

Table 12.2 provides an overview of requirements when 510(k) submission is needed and when it is not [10].

12.2.1.4 Premarket approval

The FDA has put in place an approval process for medical devices that support human life and can cause potential illness and harm to the patient. This process is called PMA and ensures sufficient controls are in place that can determine safety and effectiveness of the device. These additional controls are in the form of PMA [11].

PMA process is the scientific and regulatory review process with the FDA where the device manufacturer has to demonstrate that the device has met FDA's stringent requirements for safety and effectiveness. The technical section of the PMA application contains the data and information that will allow the FDA to determine whether to approve or disapprove the application. These sections are usually divided into nonclinical laboratory studies and clinical investigations [11]:
(a) Nonclinical laboratory studies section
   The first section is the nonclinical laboratory studies section that includes information on microbiology, toxicology, immunology,

Table 12.2 Overview of 510(k) requirements [10].

510(k) needed	510(k) not needed
1. Manufacturers who want to market and sell a device in the United States are required to make 510(k) submission at least 90 days prior to offering the device for sale 2. A change or modification to a legally marketed device and that could significantly affect its safety or effectiveness 3. A change or modifications to an existing device, where the modifications could significantly affect the safety or effectiveness of the device or the device is to be marketed for a new or different intended use 4. If devices or components sold directly to end users as replacement parts 5. If significant changes made to the labeling or otherwise affected any condition of the device	1. Unfinished devices sold to another manufacturer for further processing or selling components to be used in the assembling of devices by other manufacturers 2. Device not being marketed or commercially distributed 3. Device used for developing, evaluating, or testing 4. Devices used for clinical evaluation 5. Distributing another manufacturer's domestically manufactured device 6. If the manufacturer is the repackager or relabeler and the existing labeling condition of the device has not significantly changed 7. Commercially released device that has not significantly changed or modified in design, components, method of manufacture, or intended use 8. Importer of the foreign-made medical device and 510(k) has been submitted by the foreign manufacturer and received marketing clearance 9. Device is exempted from 510(k) by regulation (21 CFR 862-892)

Source: Food & Drug Administration, Republished on 04-Jan-2022.

biocompatibility, stress, wear, shelf life, and other laboratory or animal tests. Nonclinical studies for safety evaluation must be conducted in compliance with 21 CFR Part 58 (Good Laboratory Practice for Nonclinical Laboratory Studies).

(b) Clinical investigations section

The second section is the clinical investigations section that includes study protocols, safety and effectiveness data, adverse reactions and complications, device failures and replacements, patient information, patient complaints, tabulations of data from all individual subjects, results from statistical analysis, and any other information from the clinical investigations.

12.2.1.5 Investigational device exemption

An investigational device exemption (IDE) is a process by which an investigational device is allowed to be used in a clinical study to collect safety and effectiveness data. Most often, the clinical studies are performed to support a PMA. IDE also applies to devices that have been modified or have a new intended use of legally marketed device. All evaluations using investigational devices must have an approved IDE [12].

There are certain requirements that need to be met before devices can be evaluated [12]:
- Institutional review board has reviewed and approved investigational plan.
- Consent from all patients has been received.
- "Investigational use only" labeling affixed to the device being evaluated.
- Studies that are being performed shall be monitored.
- Records and reports are required.

Manufacturers with an approved IDE can ship a device lawfully for the purpose of conducting investigations without complying with other requirements of the Food, Drug, and Cosmetic Act. Additionally, manufacturers are not required to submit a PMA or premarket notification 510(k) while the device is under investigation.

12.2.1.6 Quality system regulation

Manufacturers must establish and follow quality systems to help ensure that their products consistently meet applicable requirements and specifications. The quality systems for FDA-regulated products (food, drugs, biologics, and devices) are known as current good manufacturing practices [13].

Quality system regulation is divided into 14 subparts:
1. Quality system requirements
2. Design controls
3. Document controls
4. Purchasing controls
5. Identification and traceability
6. Production and process controls
7. Acceptance activities
8. Nonconforming product
9. Corrective and preventive actions
10. Labeling and packaging controls
11. Handling, storage, distribution, and installation

12. Records
13. Servicing
14. Statistical techniques

Quality system requirements
This section corresponds with Chapters 4, 5, and 6 of ISO 13485.

21 CFR 820.20-25, Subpart B establishes three distinct regulatory requirements for medical device quality management systems:
- Management responsibility
- Quality audits
- Personnel

Design control
This section corresponds with Chapter 7.3.2 of ISO 13485, "Design and Development Process." Design control requirements apply to all manufacturers of class I, II, and III medical devices as well as any medical device that uses software for automation. Design control is divided into the following sections:
- Design input
- Design output
- Design review
- Design verification
- Design validation
- Design transfer
- Design changes

Design History File (DHF) and Device Master Record (DMR) are created once the design control deliverables are completed.

Document controls
This section corresponds with Chapter 4.2 of ISO 13485, which provides guidance to manufacturers for establishing and maintaining document control procedures and requirements for document approval, distribution, and document changes. Subpart D section addresses requirements for document approval, distribution, approval signatures, and access control.

Purchasing controls
This section corresponds with Chapter 7.4 of ISO 13485, which provides guidance to manufacturers for establishing purchasing controls to screen qualified suppliers and ensuring products and services meet given quality.

Identification and traceability
This section corresponds with Chapters 7.5.8 and 7.5.9 of ISO 13485, which provides guidance to manufacturers for establishing and maintaining procedures for identifying a product during every stage of the device lifecycle. Each device would require unique identifiers to ensure traceability through each stage.

Production and process controls
This section corresponds with Chapters 7.5 and 7.6 of ISO 13485, which provides guidance to manufacturers for establishing controls to ensure that devices meet the design and qualification specification. Controls should address all medical device regulatory requirements and must be documented. 820.70 subpart G lists specific requirements for equipment control and process validation, which includes documentation and calibration activities.

Acceptance activities
This section corresponds with Chapters 7 and 8 of ISO 13485, which provides guidance to manufacturers for establishing acceptance procedures for receiving, in-process, and finished product that are verified through inspection and testing. Manufacturers will develop acceptance criteria that will form the basis to accept or reject the product. All these activities are to be documented and will be part of the Device History Record.

Nonconforming product
This section corresponds with Chapter 8.3 of ISO 13485, which provides guidance to manufacturers for establishing procedures to control products that do not conform to quality specifications. Section 820.90, subpart I addresses review of nonconformance, disposition, and rework.

Corrective and preventive actions
This section corresponds with Chapter 8.5 of ISO 13485, which provides guidance to manufacturers for establishing corrective and preventive action procedures to identify quality-related issues during the device lifecycle. This includes quality issues within processes, operations, records, product return, and product complaint. Risk assessments should be performed using statistical tools to identify the root cause.

Labeling and packaging controls

This section corresponds with Chapter 7.5 of ISO 13485, which provides guidance to manufacturers for establishing controls for device labeling and inspected for label integrity. Controls should address label storage and documentation of labeling activities.

Handling, storage, distribution, and installation

This section corresponds with Chapter 7.5 of ISO 13485, which provides guidance to manufacturers for establishing controls to address safe handling, storage distribution, and installation of medical devices. Additional controls should be in place for postinstallation inspection and maintenance activities performed by patients or healthcare providers.

Records

This section corresponds with Chapters 4.2, 7, and 8 of ISO 13485, which provides guidance to manufacturers to develop and maintain comprehensive procedures for record keeping throughout the device lifecycle. These data should be made available to the FDA for review. Storage methods should be chosen that minimizes the risk of data loss. For electronic data, regular backup activities need to be performed. The control should address data confidentiality requirement and record retention.

Service

This section corresponds with Chapter 7.5.4 of ISO 13485, which provides guidance to manufacturers to establish policy for servicing, analyzing report using statistical tools and process of creating complaint procedures.

Statistical techniques

This section corresponds with all chapters of ISO 13485, which provide guidance to manufacturer to establish, control, and verify process capability and product characteristics. Section 820.250, subpart O provides guidance to establish requirements for the use of statistical tools to validate processes and products. Manufacturers must use and document appropriate sampling plan based on valid statistical method.

12.2.1.7 Labeling requirement

The FDA develops and administers regulations that apply to food, drugs, cosmetics, biologics, radiation-emitting electronic products, and medical

Table 12.3 Labeling regulations for medical devices [14].

General Device Labeling	21 CFR Part 801
General Device Labeling, Use of Symbols	21 CFR Part 801.15
In Vitro Diagnostic Products	21 CFR Part 809
Investigational Device Exemption	21 CFR Part 812
Unique Device Identification	21 CFR Part 830
General Manufacturing Practices	21 CFR Part 820
General Electronic Products	21 CFR Part 1010

Source: Food & Drug Administration, Republished on 04-Jan-2022.

devices. Labeling regulations pertaining to medical devices are found in the following parts of Title 21 of the Code of Federal Regulations (CFR) as shown below in Table 12.3 [14].

The FDA strictly enforces label guidelines as defined in Section 201(k) and 201(m) of Federal Food and Drug Cosmetic Act (FFDCA) [14].

Section 201(k) defines label as follows:

"A display of written, printed, or graphic matter upon the immediate container of any article. Any word, or statement, or other information that appears on the label shall also appear on the outside container or wrapper of the retail package or is easily legible through the outside container or wrapper".

Section 201(m) defines label as follows:

"All labels and other written, printed, or graphic matter upon any article or any of its containers or wrappers or accompanying such article".

12.2.1.8 Medical Device Reporting

Medical Device Reporting (MDR) is one of the postmarket surveillance tools the FDA uses to monitor device performance, detect potential device-related safety issues, and contribute to benefit—risk assessments of these products. It is mandatory for manufacturers, device user facilities, and importers to report adverse events and product malfunctions to the FDA about medical devices. In addition, the FDA also encourages healthcare professionals, patients, caregivers, and consumers to submit voluntary reports about serious adverse events that may be associated with a medical device, as well as use errors, and product quality issues [15].

Mandatory Medical Device Reporting requirements

The MDR regulation, 21 CFR Part 803 contains mandatory requirements for manufacturers, importers, and device user facilities to report certain device-related adverse events and product problems to the FDA [15].

Manufacturers are required to report to the FDA under the following situations [15]:
- When they learn that any of their devices may have caused or contributed to a death or serious injury
- When they become aware that their device has malfunctioned and would be likely to cause or contribute to a death or serious injury if the malfunction were to recur

Importers are required to report to the FDA under the following situations [15]:
- When importer and the manufacturer learn that one of their devices may have caused or contributed to a death or serious injury

The importer must report only to the manufacturer:
- If the imported devices have malfunctioned and would be likely to cause or contribute to a death or serious injury if the malfunction were to recur

Device user facilities must report a suspected medical device-related death to both the FDA and the manufacturer. User facilities must report a medical device-related serious injury to the manufacturer, or to the FDA if the medical device manufacturer is unknown. A device user facility refers to a hospital, surgical facility, nursing home, and outpatient diagnostic facility.

12.3 European Union Commission

The European Commission is the European Union's (EU's) politically independent executive arm that is responsible for drawing up proposals for new European legislation and implementing the decisions of the European Parliament and the Council of the EU [16].

The following are the responsibilities of European Union Commission [16]:
1. The Commission is the sole EU institution tabling laws for adoption by the Parliament and the Council that protect the interests of the EU and its citizens on issues that cannot be dealt with effectively at national level and get technical details right by consulting experts and the public.
2. The Commission sets EU spending priorities and, together with the Council and Parliament, draws up annual budgets for approval by the Parliament and Council and supervises how the money is spent, under scrutiny by the Court of Auditors.
3. The Commission enforces EU law together with the Court of Justice and ensures that EU law is properly applied in all the member countries.

4. The Commission represents the EU internationally, in particular in areas of trade policy and humanitarian aid and negotiates international agreement for the EU.

12.3.1 Medical device regulations

The European regulatory framework ensures the safety and efficacy of medical devices and facilitates patient's access to devices in the European market. Currently, medical devices in the EU are regulated by one regulation and three directives [17]:
- Regulation 2017/45 (Medical Devices)
- Council Directive 90/385/EEC (Active Implantable Medical Devices)
- Council Directive 93/42/EEC (Medical Devices)
- Directive 98/79/EC (In Vitro Diagnostic Medical Devices)

To keep up with advances in science and technology, two new regulations are replacing above three directives by 2022:
- Regulation (EU) 2017/745
- Regulation (EU) 2017/746

12.4 Other global agencies

12.4.1 European Medicines Agency

The European Medicines Agency (EMA) is a decentralized agency of the EU responsible for the scientific evaluation, supervision, and safety monitoring of medicines in the EU. EMA is governed by an independent management board. Its day-to-day operations are carried out by the EMA staff, overseen by EMA's Executive Director [18].

The mission of the EMA is to achieve scientific excellence in the evaluation and supervision of medicines, for the benefit of public and animal health in the EU [18].

The Agency uses a wide range of regulatory mechanisms to achieve the following goals:
- Support for early access
- Scientific advice and protocol assistance
- Pediatric procedures
- Scientific support for advanced therapy medicines
- Orphan designation of medicines for rare diseases
- Scientific guidelines on requirements for the quality, safety, and efficacy testing of medicines

- The Innovation Task Force, a forum for early dialogue with applicants

The Committee for Medicinal Products for Human Use (CHMP) is the EMA's subcommittee that is responsible for human medicines. The CHMP plays a vital role in the authorization of medicines in the EU.

CHMP responsibilities include the following [19]:
- Conducting the initial assessment of EU-wide marketing authorization applications
- Assessing modifications or extensions to an existing marketing authorization
- Recommending to the European Commission changes to a medicine's marketing authorization, or its suspension or withdrawal from the market

The CHMP also evaluates medicines authorized at national level referred to EMA for a harmonized position across the EU. In addition, the CHMP and its working parties contribute to the development of medicines and medicine regulation, by
- providing scientific advice to companies researching and developing new medicines;
- preparing scientific guidelines and regulatory guidance to help pharmaceutical companies prepare marketing authorization applications for human medicines; and
- cooperating with international partners on the harmonization of regulatory requirements.

12.4.2 National Medical Products Administration, China

The National Medical Products Administration (NMPA) is the Chinese agency for regulating drugs and medical devices. It was formerly known as the China Food and Drug Administration (CFDA) [20].

NMPA responsibilities include the following [20]:
- Draft laws and regulations for drugs, medical devices, and cosmetics, as well as establishing medical device standards and classification systems.
- Develop good practices on the distribution and use of medical products.
- Develop good manufacturing practices (GMP), good laboratory practices (GLP), and good clinical practices (GCP).
- Undertake quality management for drugs, medical devices, and cosmetics.
- Regulate the registration of drugs, medical devices, and cosmetics.
- Postmarket risk management for drugs, medical devices, and cosmetics.

- Organize the monitoring, evaluation, and handling of adverse drug reactions, medical device adverse events, and cosmetic adverse reactions.

The NMPA departments dealing with medical devices are as follows:
- Department of Medical Device Registration, responsible for premarket approvals
- Department of Medical Device Supervision, responsible for postmarket supervision

Medical device approval is required to be obtained from NMPA. The premarket or product registration procedures are different for imported products and local domestic products. For foreign manufacturers who are importing their medical devices into China, a product registration license will be required. For domestic products, both the product registration license and production approval license are mandatory [21].

Three Chinese regulatory agencies, namely, the NMPA, the provincial bureau, and the municipal bureau oversee medical device approval process. Table 12.4 shows details of medical device approval process [21].

For medical devices of class I, foreign manufacturers are responsible for one kind of record, and domestic manufacturers are responsible for two kinds of records. For class II and class III medical devices, foreign manufacturers will obtain one license, which is the medical devices registration certificate. Domestic manufacturers will obtain two licenses, which are the medical devices registration and the production license [21].

Table 12.4 Details of medical device approval process.

	Class I	Class II	Class III
Imported products	NMPA Record of imported medical devices	NMPA Medical devices registration certificate	NMPA Medical devices registration certificate
Local registrant	NMPA Market authorization holder license	NMPA Market authorization holder license	NMPA Market authorization holder license
Domestic products	Municipal bureau Record of domestic medical devices	Provincial bureau Medical devices registration certificate	NAMPA Medical devices registration certificate
Domestic manufacturers	Municipal bureau Record of production	Provincial bureau Production license	Provincial bureau Production license

Foreign manufacturers are required to have a local representative company to be the marketing authorization holder (MAH) for the submission of their medical device registration. There may be some special situations in which the MAH requirement is applicable to domestic manufacturers [21].

12.4.3 Central Drugs and Standard Control Organization, India

The Central Drugs and Standard Control Organization (CDSCO) is the Indian regulatory body for pharmaceuticals and medical devices. The CDSCO is responsible for regulating manufacture, approval, and sale of medical devices and drugs in India. They are also responsible for approving clinical trials as well as providing expert advice on health issues and enforcement of the Drugs and Cosmetic Act [22,23].

Drugs and Cosmetic Act cover wide varieties of therapeutic substances, diagnostics, and medical devices [23]. The department within CDSCO responsible for regulating medical devices and drugs is known as the Drug Controller General of India (DCGI). Inspections, audits, postmarket surveillance, and recalls are done through the department that is split into zonal offices. Although limited number of medical devices and IVDs require registration in India, foreign medical device manufacturers are required to comply with India's medical device regulations before they can sell in India [22].

The first step for manufacturers is to appoint an Indian authorized agent who would interact with the CDSCO on their behalf. If the device requiring approval is a class B, C, or D IVD, the manufacturer would need in-country performance testing done through the National Institute of Biologicals (NIB).

Next step is to compile device application, Form MD-14 including manufacturing facility information, device technical information, ISO 13485 certificate, IFU, testing results (if applicable), clinical data (if applicable), proof of approval in the United States, EU, Australia, Canada, or Japan, plus proof of approval in manufacturers home country. Application for registration/Import License is then filed with the CDSCO and fees paid [22].

The CDSCO then reviews the application and in some cases may require technical presentation before granting an approval. Upon completion of the review, the CDSCO issues an import license in Form MD-15.

12.4.4 Health Canada

Health Canada is the Canadian regulatory agency for pharmaceuticals and medical devices. It is responsible for regulating manufacture, approval, and sale of medical devices and drugs in Canada. To manufacture, market, and sell product in Canada, manufacturers are required to obtain a license. Health Canada issues two types of licenses—(1) Medical Device Establishment License (MDEL) and (2) Medical Device License (MDL) [24].

The first step of the process is to determine the classification of the medical device. Except for class I devices, manufacturers are required to implement an ISO 13485 under Medical Device Single Audit Program (MDSAP) compliant quality management system. Next step is to have ISO 13485 quality system audited by an approved auditing organization under MDSAP. A new ISO 13485 certificate is issued upon successful completion of the audit. Manufacturer applies for MDEL if it is a class I device and MDL if it is class II, III, and IV devices. Submit the application with Health Canada and pay applicable fees [24].

Health Canada reviews the application along with premarket review document. For class I devices, approved applications are posted on the Health Canada website and MDEL certificate emailed to the manufacturer. For class II, III, and IV devices, issued licenses are posted on the Health Canada website, and copies of the MDL emailed to the manufacturer [24]. At this stage, the manufacturer can start marketing their device in Canada.

12.5 International Medical Device Regulators Forum

The International Medical Device Regulators Forum (IMDRF) is a voluntary group of medical device regulators from across the world who have come together to create a Global Harmonization Task Force on Medical Devices. The goal of this task force is to accelerate international medical device regulatory harmonization [25]. IMDRF management committee provides guidance on strategies, policies, directions, membership, and activities of the forum.

Working groups are established by the management committee and operate under targeted mandates on specific activities. Some of the current work items include the following [26]:
- Artificial intelligence medical devices
- Principles of in vitro diagnostic

- Medical devices classification
- Medical device cybersecurity guide
- Medical device clinical evaluation
- Personalized medical devices
- Adverse event terminology
- Good regulatory review practices
- Regulated product submission

References

[1] Food and Drug Administration. https://www.fda.gov/regulatory-information/laws-enforced-fda.

[2] Medical Devices: Full Definitions, World Health Organization, Geneva. https://www.who.int/medical_devices/full_deffinition/en/.

[3] L.R. Atles, A Practicum for Biomedical Engineering and Technology Management Issues, Kendall Hunt Publishing, Dubuque, 2008.

[4] Medical Device Timeline, Morgridge Institute for Research, Madison, WI. https://morgridge.org/outreach/teaching-resources/medical-devices/medical-devices-timeline/.

[5] R. Fenton, What are the Differences in the FDA Medical Device Classes, Qualio Blog, Quality & Compliance Hub, June 2021. https://www.qualio.com/blog/fda-medical-device-classes-differences.

[6] Food and Drug Administration, Classifying Medical Devices. https://www.fda.gov/medical-devices/overview-device-regulation/classify-your-medical-device.

[7] Food and Drug Administration, Overview of Medical Device Classification and Reclassification. https://www.fda.gov/about-fda/cdrh-transparency/overview-medical-device-classification-and-reclassification.

[8] Food and Drug Administration, Code of Federal Regulations. https://www.fda.gov/medical-devices/overview-device-regulation/code-federal-regulations-cfr.

[9] Food and Drug Administration, Device Registration and Listing. https://www.fda.gov/medical-devices/how-study-and-market-your-device/device-registration-and-listing.

[10] Food and Drug Administration, Premarket Notification 510(k). https://www.fda.gov/medical-devices/premarket-submissions/premarket-notification-510k.

[11] Food and Drug Administration, Premarket Approval (PMA). https://www.fda.gov/medical-devices/premarket-submissions/premarket-approval-pma.

[12] Food and Drug Administration, Investigational Device Exemption. https://www.fda.gov/medical-devices/how-study-and-market-your-device/investigational-device-exemption-ide.

[13] Food and Drug Administration, Quality System Regulations. https://www.fda.gov/medical-devices/postmarket-requirements-devices/quality-system-qs-regulationmedical-device-good-manufacturing-practices.

[14] Food and Drug Administration, Device Labeling. https://www.fda.gov/medical-devices/overview-device-regulation/device-labeling.

[15] Food and Drug Administration, Medical Device Reporting. https://www.fda.gov/medical-devices/medical-device-safety/medical-device-reporting-mdr-how-report-medical-device-problems#overview.

[16] European Union Official Website. https://europa.eu/european-union/about-eu/institutions-bodies/european-commission_en.

[17] European Commission, Medical Devices Sector. https://ec.europa.eu/health/md_sector/overview_en.
[18] European Medicines Agency. https://www.ema.europa.eu/en/.
[19] Committee for Medicinal Products for Human Use (CHMP). https://www.ema.europa.eu/en/committees/committee-medicinal-products-human-use-chmp.
[20] National Medical Products Administration. http://english.nmpa.gov.cn/index.html.
[21] M. Ng, R. Dazhi, A Road Map to China's Medical Device Registration Process, Med Device Online, April 2021, in: https://www.meddeviceonline.com/doc/a-road-map-to-china-s-medical-device-registration-process-0001.
[22] Central Drugs Standard Control Organization. https://cdsco.gov.in/opencms/opencms/en/Home/.
[23] Drugs and Cosmetic Act, Government of India. https://www.nhp.gov.in/drugs-and-cosmetics-act_mtl.
[24] Safe Medical Devices in Canada, Health Canada. https://www.canada.ca/en/health-canada/services/drugs-health-products/medical-devices/activities/fact-sheets/safe-medical-devices-fact-sheet.html.
[25] International Medical Device Regulators Forum. http://www.imdrf.org/about/about.asp.
[26] Work Items, International Medical Device Regulator Forum. http://www.imdrf.org/workitems/work.asp.

CHAPTER THIRTEEN

Market trends and global sourcing of medical devices

Contents

13.1 Market trends toward increased safety and quality of medical devices	241
13.2 Market trends in the medical device industry	243
13.2.1 Medical devices market drivers	243
13.2.2 Medical device market restraints	244
13.2.3 Medical devices market trends	244
13.2.3.1 Expansion of medical technologies	244
13.2.3.2 Self-diagnosis/treatment is increasing	244
13.3 Global sourcing of plastics	245
13.3.1 Quality assurance	245
13.3.1.1 Quality assurance framework for procurement of medical device	245
References	248

13.1 Market trends toward increased safety and quality of medical devices

Rapid globalization means that companies need to do everything they can to remain competitive. Increased pressure on medical device companies means they are constantly looking to reduce costs and improve standards. Quality issues can be the result of a slip somewhere along a complex supply chain or during the design, testing and manufacturing process. A single error can have severe consequences not just for patients, but shareholders too: one major quality event can mean a 10% drop in a manufacturer's share price [1].

Companies are also aware of the risk. According to research, 67% of executives consider cost of quality essential to competitive success. At the same time, the majority of manufacturers rely on cost of quality to boost customer satisfaction (89%) and to remain ISO 9000 compliant (84.5%) [2].

Further research by McKinsey reveals that major recalls or other quality events can mean medical device manufacturers risk losing up to 11.7% of market segment revenue or around $300 million in less than 12 months [3].

McKinsey identified sources of maturity that relate to good quality in medical devices [3]:

- By linking critical quality attributes (CQAs) with critical control points (CCPs) during the production and manufacturing processes, companies can achieve more robust product and process controls.
- Stronger operational maturity relating to people and assets means a greater focus on employee retention and shared targets to help manufacturers improve quality.
- Mature quality systems—specifically, supplier controls and nonconformance, corrective action/preventive action (CAPA) management—drive better quality performance and reduce quality cost.
- Implementing an analytics-driven quality culture and processes across an organization is also critically important

The FDA's Center for Devices and Radiological Health (CDRH) has developed "Medical Device Safety Action Plan" that focuses on five key elements [4]:

- Establish a robust medical device patient safety net in the United States.
- Explore regulatory options to streamline and modernize timely implementation of postmarket mitigations.
- Spur innovation toward safer medical devices.
- Advance medical device cybersecurity.
- Integrate CDRH's premarket and postmarket offices and activities to advance the use of a total product life cycle (TPLC) approach to device safety.

The objective of this plan is to ensure that medical devices meet the gold standard to get to market and provide clinical data highlighting benefit—risk about the device and a pathway for medical device manufacturer to expedite the process of demonstrating safety and efficiency for moderate risk devices. These developments have impacted areas like machine learning diagnostic tools, wearables, and digital health technology, among others [4].

The FDA has increased cooperation with medical providers and developers to streamline the approval process for new technology. With benefits for both medical device companies and the patients who use their products, these reduced regulations indicate the FDA's commitment to innovation and progress [4].

13.2 Market trends in the medical device industry

The medical device industry includes corporations that manufacture medical devices. The market is typically segmented by the following:
- Type of device
- Type of expenditure
- End user
- Geography

Medical devices based on the type of device include the following [5]:
- In vitro diagnostics
- Dental equipment and supplies
- Ophthalmic devices
- Diagnostic imaging equipment
- Cardiovascular devices
- Hospital supplies
- Surgical equipment
- Orthopedic devices
- Patient monitoring devices
- Diabetes care devices
- Nephrology and urology devices
- ENT devices
- Anesthesia and respiratory devices
- Neurology devices
- Wound care devices

According to market research company, The Business Research Company, the global medical devices market reached a value of nearly $456.9 billion in 2019, having increased at a compound annual growth rate (CAGR) of 4.4% since 2015. The market is expected to decline from $456.9 billion in 2019 to $442.5 billion in 2020 at a rate of −3.2% primarily due to successive lockdowns due COVID-19 pandemic. Due to the increased demand for ventilators, the market is expected to grow at a CAGR of 6.1% from 2021 and reach $603.5 billion in 2023 [5−10].

13.2.1 Medical devices market drivers

The main driver of the medical devices market is the increasing prevalence of chronic diseases such as diabetes and cancer. These diseases are expected to be a major driver of the medical devices market. The proportion of total global deaths due to chronic diseases is expected to increase to 70%, and the global burden of chronic diseases is expected to reach about 60% by 2030 as reported by the United Nations [5].

The major causes of chronic diseases are as follows:
- Extended working hours
- Limited physical activity
- Unhealthy food habits

13.2.2 Medical device market restraints

The main restraint on the medical devices market is the data security issue. More and more people are adopting to use of remote patient monitoring devices, which has increased the threats posed by data security issues. Remote patient monitoring devices use the Internet to send a patient's data to a physician. The connectivity of the devices to the global networks makes them targets for hacks and data thefts, which can cause leak of valuable and sensitive patients' information [5].

13.2.3 Medical devices market trends

There are two major trends that influence medical devices market:
- Expansion of medical technologies
- Increase in self-diagnosis

13.2.3.1 Expansion of medical technologies

Medical technologies such as wearables are becoming more commonly used for diagnosing, treating, and monitoring patients' health without human contact. These technologies rely on artificial intelligence to provide contactless monitoring of patients in their home or in situations where access to the patient is limited. Examples of such technologies include wearable medical equipment, remote patient monitoring devices, and electronic health records (EHRs) [5].

13.2.3.2 Self-diagnosis/treatment is increasing

Technological developments in devices such as glucose monitors, insulin delivery devices, nebulizers, and oxygen concentrators have enabled diagnosis and monitoring of many diseases at home. More and more patients are using home-based medical devices for diagnosis and treatment of medical conditions. Remote control technology is also allowing healthcare professionals to support home-based treatments, which is leading to increased preference for home and self-care treatment [5].

13.3 Global sourcing of plastics
13.3.1 Quality assurance

Quality assurance (QA) is part of the manufacturing system that monitors implementation of the quality system and influence the quality of a product. The overall objective is to ensure that products meet the quality standard. QA incorporates several factors and is an integral part of all key activities in the product supply chain. Factors that define product quality are as follows [11]:

- Raw materials
- Quality control
- Manufacturing process
- Packaging and labeling
- Regulatory approval
- Transportation and distribution
- Storage

13.3.1.1 Quality assurance framework for procurement of medical device

QA framework for the procurement of medical devices comprises the following:

- Prequalification
- Purchasing
- Receipt and storage
- Distribution

Prequalification

Prequalification is the first step of the quality assurance framework for procurement of medical devices. During prequalification, the following activities are performed:

- Define the need of a product.
- Assess the product offered against the specifications.
- Assess the facility where the product is manufactured against cGMP standards.

Key steps in the prequalification process:
1. Solicit information.
2. Receive product information.
3. Screen product information.

4. Evaluate product information.
5. Perform inspection.
6. Finalize assessment process.

Solicit information

Product specifications Product specifications need to be developed for list of products with information that would assist in documentation. The specification may include following information:
- Material
- Primary packaging materials
- Lot size
- Shelf life
- Labeling requirements

Establish quantification Quantities for each product should be specified when a request is made. To avoid shortages or excess inventory, accurate forecast and supply plan should be developed. A reliable estimate of actual need should be used when purchasing given quantities.

Define procurement method Based on organizational policy and procedures, the procurement method should be identified, defined, and used. Some of the procurement methods are shown in the following:
- Restricted tender
- Competitive negotiation
- Direct procurement
- Open tender

Establish submission procedure The procedure should be clearly written and contain information on the following:
- Content and format of submission
- Process of submission

Receive product information

The procurement agency should have the necessary infrastructure to receive and process the product information package submitted by manufacturers. It will require personnel for processing the documentation; written procedures for receiving, identifying, and marking files, containers, and samples; and sufficient space for unpacking and storage.

Each product should be allocated a unique reference number to ensure traceability of the product information package. A record of all the information received from each manufacturer should be maintained.

Screen product information
Each product information package submitted by the manufacturer should be screened for completeness. Information to be recorded should include the following:
- Date of receipt
- Product number
- Name of product
- Name of the applicant (i.e., supplier)
- Name and address of manufacturer
- Outcome of screening

Evaluate product information
A qualified person with relevant qualifications and experience should evaluate product information. Each evaluator should prepare a formal evaluation report for each product, including a recommendation for acceptance or rejection. The evaluation report should be communicated to the manufacturer. A response should be requested from the manufacturer in situations where data and information are found to be incomplete or do not meet the guidelines. An evaluation report should be filed with product documentation.

Perform inspection
Inspection should be performed in accordance with a written procedure to ensure all aspects of cGMP for the relevant product(s) have been covered. Any report of incidents with the product or the manufacturer should be reported. The inspection should cover the evaluation and assessment of the manufacturing documentation, premises, equipment, utilities, and materials. It should also cover verification of data and documentation, such as results, batch records, compliance with an SOP, as well as information submitted on the manufacturing method, equipment. Inspection should further include validation of the manufacturing process, validation of utilities and support systems, and validation of equipment.

Finalize assessment process
Assessment should be made based on evaluation of the product, cGMP compliance, and test results. Decision to accept or reject a product and/or manufacturer should be made based on this assessment.

Purchasing
Any purchase that needs to happen should not primarily focus on price but needs to be performed with the aim of purchasing effective, safe, and quality products. Prequalified products are purchased from approved suppliers.

Metrics need to be developed for continuous monitoring of the performance of manufacturers and suppliers. This may be a joint responsibility of the QA personnel and the purchasing group. Monitoring may include the following:
- Review of quality control test results
- Verification that product batches supplied have been manufactured in compliance with standards and specifications accepted in the product dossier through inspection
- Review of rejected or failure batches
- Monitoring of complaints and recall
- Outcome of reinspection of manufacturing sites
- Outcome of reevaluation of product information
- Monitoring of direct and indirect product costs
- Monitoring of adherence to delivery schedules

Receipt and storage

Medical devices purchased should be received and stored correctly and in compliance with good storage practices and good documentation practices and follow all applicable local and federal regulations. A well setup receipt and storage should maintain high standards of quality and integrity. Batch records should be traceable, and stocks should be allowed to be rotated. There should be a unidirectional flow of products, i.e., from receiving to dispatch to avoid potential mix-ups.

Distribution

The objectives of well-maintained distribution center are as follows:
- Maintain a constant supply of raw materials.
- Maintain accurate inventory records.
- Use available transportation resources as efficiently as possible.
- Reduce theft and fraud.
- Provide information for forecasting device needs.

References

[1] L. Ljungberg, The importance of meeting quality requirements in the medtech industry, ELOS MEDTECH. https://elosmedtech.com/the-importance-of-meeting-quality-requirements-in-the-medtech-industry/.

[2] P. Kaur, Current cost of quality management practices in India in the era of globalization: an empirical study of selected companies, Decision 36 (1) (2009).

[3] E. Makarova, Capturing the value of good quality in medical devices, McKinsey and Company. https://www.mckinsey.com/industries/pharmaceuticals-and-medical-products/our-insights/capturing-the-value-of-good-quality-in-medical-devices.

[4] J. Catley, What the medical device safety action plan means for medical marketers, Medical Marketing Insights, 2018, https://www.mdconnectinc.com/medical-marketing-insights/medical-device-safety-action-plan.
[5] Medical Device Global Market Opportunities and Strategies, The Business Research Company. https://www.thebusinessresearchcompany.com/report/medical-devices-market.
[6] Medical Device Market — Forecast (2021—2026), Industry Arc. https://www.industryarc.com/Research/Medical-Device-Market-Research-514107.
[7] Medical Device Market Research Report, Fortune Business Insights. https://www.fortunebusinessinsights.com/industry-reports/medical-devices-market-100085.
[8] Medical Device Market Report: Trends, Forecast and Competitive Analysis, Lucintel. https://www.lucintel.com/medical-device-market.aspx.
[9] W. Kluwer, Medical Device Outlook for 2021 and Beyond. https://www.wolterskluwer.com/en/expert-insights/medical-device-outlook.
[10] Precedence Research, Medical Devices Market Size to Hit Around US$671.49Bn by 2027, 2021, https://www.globenewswire.com/news-release/2021/06/30/2255807/0/en/Medical-Devices-Market-Size-to-Hit-Around-US-671-49-Bn-by-2027.html.
[11] USAID, Module 1: Quality Assurance in Procurement, Manual for Procurement & Supply for Quality-Assured MNCH Commodities. https://www.ghsupplychain.org/node/615.

CHAPTER FOURTEEN

Reprocessing of reusable medical devices

Contents

14.1 Overview—reusable medical devices	251
14.2 FDA guidance on reprocessing	252
14.3 Factors affecting quality of reprocessing	253
14.3.1 Device design	253
14.3.2 Reprocessing methodology	254
14.3.3 Methods for validating cleaning and high-level disinfection or sterilization instructions	254
14.4 Cleaning validation—reusable medical devices	255
14.5 Validation of the final microbicidal process	256
14.6 Reusable medical device labeling	256
14.7 Reprocessing of single-use medical devices	258
References	259

14.1 Overview—reusable medical devices

Reusable medical devices are devices that healthcare providers can reprocess and reuse on multiple patients. Examples of reusable medical devices include surgical forceps, endoscopes, and stethoscopes [1].

All reusable medical devices can be grouped into one of three categories according to the degree of risk of infection associated with the use of the device:

- Critical devices, such as surgical forceps, come in contact with blood or normally sterile tissue.
- Semicritical devices, such as endoscopes, come in contact with mucus membranes.
- Noncritical devices, such as stethoscopes, come in contact with unbroken skin.

Critical and semicritical devices are designed and labeled for multiple uses and are reprocessed by thorough cleaning followed by high-level disinfection or sterilization between patients. They are made of materials that can withstand repeated reprocessing, including manual brushing and the use of chemicals.

Some examples of reusable medical devices are as follows:
- Surgical instruments, such as clamps and forceps
- Endoscopes, such as bronchoscopes, duodenoscopes, and colonoscopes, used to visualize areas inside the body
- Accessories to endoscopes, such as graspers and scissors
- Laparoscopic surgery accessories, such as arthroscopic shavers

14.2 FDA guidance on reprocessing

Reusable medical devices such as surgical forceps, endoscopes, and stethoscopes when used on patients become soiled and contaminated with microorganisms. To avoid any risk of infection by a contaminated device, reusable device needs to be cleaned and sterilized. A detailed, multistep process to clean and then disinfect or sterilize medical devices is termed as reprocessing. Reprocessed medical devices can be safely used more than once in the same patient, or in more than one patient if the labeling instructions for reprocessing are completely and correctly followed after each use of the device [2].

Inadequate cleaning between patient uses can result in the retention of blood, tissue, and other biological debris in certain types of reusable medical devices. This debris can allow microbes to survive the subsequent disinfection or sterilization process, which could then lead to healthcare-associated infections (HAIs).

Inadequate reprocessing can also result in other adverse patient outcomes such as tissue irritation from residual reprocessing materials, such as chemical disinfectants. Infections from inadequately reprocessed devices are not often recognized or reported to the FDA. The number of HAIs that can be attributed to inadequate device reprocessing is unknown because it is not often investigated as a cause of HAI [2].

The FDA published a final guidance on reprocessing reusable medical devices. This guidance is a step toward further reducing the risk of patient infection by providing manufacturers with recommendations to validate their reprocessing instructions to ensure devices remain safe and effective for reuse [3,4].

The final guidance document includes the following [3,4]:
- The scientific advances in knowledge and technology involved in reprocessing reusable medical devices

- General considerations for reusable medical device design and reprocessing instructions in device labeling
- Six specific criteria that manufacturers should address in reprocessing instructions
  - Criterion 1: Labeling should reflect the intended use of the device
  - Criterion 2: Reprocessing instructions for reusable devices should advise users to thoroughly clean the device
  - Criterion 3: Reprocessing instructions should indicate the appropriate microbicidal process for the device
  - Criterion 4: Reprocessing instructions should be technically feasible and include only devices and accessories that are legally marketed
  - Criterion 5: Reprocessing instructions should be comprehensive
  - Criterion 6: Reprocessing instructions should be understandable
- Recommended reprocessing validation methods designed to clean, disinfect, and sterilize reusable medical devices
- A subset of medical devices for which 510(k) submissions should include protocols and complete test reports of the validation of the reprocessing instructions to demonstrate that reprocessing methods and instructions are adequate

14.3 Factors affecting quality of reprocessing

Quality of reprocessing can be affected by three factors [5]:
- Device design
- Reprocessing methodology
- Methods for validating the cleaning and high-level sterilization instructions

14.3.1 Device design

Due to complex nature of designs of some types of reusable medical devices, it becomes difficult to achieve optimal cleaning and high-level disinfection or sterilization. The FDA has identified design features that are prone to retain debris and biological materials based on its evaluation of adverse event reports and include the following [5]:
- Long, narrow interior channels, including those with internal surfaces that are not smooth, have ridges or sharp angles, or are too small to permit a brush to pass-through
- Hinges
- Sleeves surrounding rods, blades, activators, inserters, and so on

- Adjacent device surfaces between which debris can be forced or caught during use
- O-rings
- Valves that regulate the flow of fluid through a device
- Devices with these or other design features that cannot be disassembled for reprocessing

14.3.2 Reprocessing methodology

Reprocessing is detailed, labor intensive, and time-consuming and can be prone to errors. Each reusable medical device requires specific reprocessing steps or techniques appropriate for that device [5].

Many variables impact the effectiveness of reprocessing reusable medical devices:
- Reprocessing challenges at individual facilities, such as follows:
  - Staff responsible for steps in the process
  - Training available to the staff
  - Equipment (e.g., appropriately sized brushes) available for use
- Quality and completeness of the reprocessing instructions provided by the manufacturer
- Access to the manufacturer's instructions

14.3.3 Methods for validating cleaning and high-level disinfection or sterilization instructions

Manufacturers are required to validate their reprocessing instructions by documenting that the recommended cleaning and sterilization process consistently results in an adequately reprocessed device.

Cleaning validation involves three important steps [5]:
- Soiling of the devices from clinical use, or simulated soiling with a test soil
- Cleaning of the device
- Identifying a method to measure residual components of the test soil remaining on the device

Based on the assessed data, the FDA has identified top two reasons that can impact quality of medical device reprocessing:
- Designing inadequate test conditions
- Use of inappropriate test methods to validate

14.4 Cleaning validation—reusable medical devices

Reprocessing of medical devices involves cleaning, disinfection, and sterilization. Cleaning validation should demonstrate the following [6]:
- Cleaning methods specified are adequate for the device to undergo further processing and reuse.
- Reprocessing instructions are effective in conveying the proper reprocessing methods to the user.

Cleaning validation activities should use the worst-case testing method using soils that are relevant to the clinical use condition of the device. The worst-case means that the worst-case implementation of the cleaning process should be used, i.e., the least rigorous cleaning process. Medical devices representing worst-case, i.e., the device should be the most challenging to reprocess and should be the most contaminated, should be used in the validation process. The FDA recommends that at least two quantitative test methods that are related to the clinically relevant soil should be used [6].

The cleaning process validation protocols should specify predetermined cleaning test endpoints. The protocols should be designed to establish that the most inaccessible locations on your devices can be adequately cleaned during routine processing. The validation protocols should select artificial soil test which composition represents the materials that the device is likely to be exposed to during actual clinical use and especially for the worst-case cleaning challenge. Soil inoculations should mimic worst-case clinical use conditions. Validation studies should include simulated use conditions, especially for devices with features at risk for the accumulation of soil with repeated use [6].

Validation protocols should support cleaning instructions and should be detailed and specific with respect to parameters such as time, temperature, and concentrations. For choosing the two FDA recommended quantitative test methods, factors that should be considered include the following:

(a) contaminants the device is expected to come in contact with during actual use
(b) test specificity for direct measurement of those constituents
(c) sensitivity of test methods in relation to proposed cleaning endpoints

Test methods used to measure residual soil should be validated. Analytical sensitivity and specificity information, predetermined cleaning endpoints, and appropriate controls should be documented. Devices should be subjected to a validated method of extraction for recovery of residual soil. The extraction method should be completely described for each device, and its recovery efficiency should be determined as part of its validation [6].

14.5 Validation of the final microbicidal process

The FDA recommends that the disinfection processes and instructions should be validated and the recommendations in device-specific FDA guidance documents or any relevant FDA-recognized standards should be followed [6].

For reusable devices that are intended to be used sterile, labeling should include a sterilization process that has been validated to attain a sterility assurance level (SAL) of 10^{-6} (or 10^{-3}, as appropriate). Validation data should be generated in FDA-cleared sterilizers and with FDA-cleared sterilization accessories such as biological indicators, physical/chemical sterilization process indicators, and sterilization wraps.

The second approach is to have validation data generated in sterilizers that can show equivalent or better control of key sterilization parameters than FDA-cleared sterilizers. If the second approach is chosen, differences that may exist between the test sterilizer and the FDA-cleared sterilizer should be addressed [6].

14.6 Reusable medical device labeling

The FDA addresses the following six criteria for clear reprocessing instructions. They are as follows:
- Criterion 1: Labeling should reflect the intended use of the device.
- Criterion 2: Reprocessing instructions for reusable devices should advise users to thoroughly clean the device.
- Criterion 3: Reprocessing instructions should indicate the appropriate microbicidal process for the device.
- Criterion 4: Reprocessing instructions should be technically feasible and include only devices and accessories that are legally marketed.
- Criterion 5: Reprocessing instructions should be comprehensive.
- Criterion 6: Reprocessing instructions should be understandable.

(a) Criterion 1: Labeling should reflect the intended use of the device.

The FDA recommends that labels should include instructions for a reprocessing method that reflects the physical design of the device, its intended use, and the soiling and contamination to which the device will be subjected during clinical trials [6].

(b) Criterion 2: Reprocessing instructions for reusable devices should advise users to thoroughly clean the device.

The FDA recommends that cleaning should be described in the labeling as part of the overall reprocessing instructions since it is the first step in reprocessing. If there are directions for use of device that may include the use of protective covers, labels should include the recommendation to use only legally marketed protective covers [6].

(c) Criterion 3: Reprocessing instructions should indicate the appropriate microbicidal process for the device.

In some clinical situations such as patients with Norovirus infections or drug-resistant organisms, isolation precautions recommended for use by the Center for Disease Control and Prevention (CDC) may include the use or specific disinfectants and should be followed. The FDA recommends label to instruct a user to follow the specific EPA label disinfectant contact times when using the disinfectant as well as the instructions specified in the medical device label [6].

(d) Criterion 4: Reprocessing instructions should be technically feasible and include only devices and accessories that are legally marketed.

The FDA recommends that the instructions in the label that specify sterilization methods and parameters are to be technically feasible. Sterilization cycle parameters specified in the labeling for reprocessing a device should be consistent with validated sterilization cycle parameters available in legally marketed sterilizers [6].

(e) Criterion 5: Reprocessing instructions should be comprehensive.

Labeling should include applicable instructions for point-of-use processing. For disassembly and reassembly, the labeling should provide with a validated method to verify that reassembly has been properly performed to assure that the device is in operable condition for the next use [6].

For method of cleaning, labeling should contain comprehensive directions, including photographs and/or diagrams, if appropriate, for each cleaning, rinsing, and drying step so that users can accurately follow the steps or program them into the device washer or washer/disinfector. Recommendations for the use of detergents, enzymatic cleaners, and automated cleaning cycles should be consistent with the manufacturer's directions for use for those products [6].

14.7 Reprocessing of single-use medical devices

Reprocessing single-use devices (SUDs) involves reusing devices that were designed and sold for single-use only. To save money, hospitals are moving toward using third-party reprocessed SUDs. However, reusing devices intended for single use can be dangerous without the correct validations and instructions for reprocessing in place. SUDs that were not designed to be effectively cleaned and resterilized may contain areas not accessible to thorough cleaning and may fail to withstand the harsh conditions (e.g., exposure to solvents and extreme temperatures) encountered during reprocessing [7–9].

Postmarket validations need to occur to reuse devices intended for single use. Since there are no reprocessing instructions for use (IFU) for SUDs, cleaning and sterilization processes must be developed and validated to ensure patient safety.

The FDA released a guidance document on SUDs reprocessed by third parties or hospitals. In this guidance document, the FDA states that hospitals or third-party reprocessors will be considered manufacturers and regulated in the same manner. A reused SUD will have to comply with the same regulatory requirements of the device when it was originally manufactured. This document presents FDA's intent to enforce premarket submission requirements within 6 months for class III devices, 12 months for class II devices, and 18 months for class I devices [7–9].

The FDA uses two types of premarket requirements for nonexempt class I and II devices, a 510(k) submission that may have to show that the device is as safe and effective as the same device when new, and a premarket approval application. The 510(k) submission must provide scientific evidence that the device is safe and effective for its intended use.

The FDA allows hospitals a year to comply with the nonpremarket requirements (registration and listing, reporting adverse events associated with medical devices, quality system regulations, and proper labeling). The options for hospitals are to stop reprocessing SUDs, comply with the rule, or outsource to a third-party reprocessor. FDA guidance document does not apply to permanently implantable pacemakers, hemodialyzers, opened but unused SUDs, or healthcare settings other than acute-care hospitals [7–9].

References

[1] Food and Drug Administration. What are Reusable Medical Devices. https://www.fda.gov/medical-devices/reprocessing-reusable-medical-devices/what-are-reusable-medical-devices.

[2] Food and Drug Administration. Reprocessing of Reusable Medical Devices. https://www.fda.gov/medical-devices/products-and-medical-procedures/reprocessing-reusable-medical-devices.

[3] Food and Drug Administration. Reprocessing of Reusable Medical Devices: Information for Manufacturer. https://www.fda.gov/medical-devices/device-advice-comprehensive-regulatory-assistance/reprocessing-reusable-medical-devices-information-manufacturers.

[4] Food and Drug Administration. https://www.fda.gov/medical-devices/reprocessing-reusable-medical-devices/working-together-improve-reusable-medical-device-reprocessing.

[5] Food and Drug Administration. Factors Affecting Quality of Reprocessing. https://www.fda.gov/medical-devices/reprocessing-reusable-medical-devices/factors-affecting-quality-reprocessing.

[6] Food and Drug Administration, Reprocessing Medical Devices in Health Care Settings: Validation Methods and Labeling, Guidance for Industry and Food and Drug Administration Staff, Docket Number FDA-2011-D-0293, March 2015, https://www.fda.gov/regulatory-information/search-fda-guidance-documents/reprocessing-medical-devices-health-care-settings-validation-methods-and-labeling.

[7] L. Bookoff, FDA Updates Guidance on Reprocessing of Single-Use Devices, Medical Device and Diagnostic Industry, October 2006, in: https://www.mddionline.com/news/fda-updates-guidance-reprocessing-single-use-devices.

[8] Reprocessed Single-Use Devices, The American College of Obstetricians and Gynecologists, March 2019. Number 769, https://www.acog.org/clinical/clinical-guidance/committee-opinion/articles/2019/03/reprocessed-single-use-devices.

[9] Reuse of Single-Use Medical Devices. Center for Disease Control and Prevention. https://www.cdc.gov/infectioncontrol/guidelines/disinfection/reuse-of-devices.html.

CHAPTER FIFTEEN

Economics and future direction of medical devices

Contents

15.1 Understanding the economics of medical device cost	261
15.1.1 Material	262
15.1.2 Fabrication	263
15.1.2.1 Material	*263*
15.1.2.2 Labor	*263*
15.1.2.3 Part design and accuracy	*263*
15.1.2.4 Finishing and quantity	*264*
15.1.3 Quality	264
15.1.3.1 Direct costs of ensuring good quality	*264*
15.1.3.2 Direct costs of poor quality	*265*
15.1.3.3 Indirect quality costs	*266*
15.1.4 Validation	266
15.1.4.1 Design validation	*267*
15.1.4.2 Process Validation	*268*
15.2 Migration of device manufacturing, validation and use	270
15.2.1 Consideration for manufacturing transfer	270
15.2.1.1 Planning	*270*
15.2.1.2 Consistent communication	*270*
15.2.1.3 Materials	*271*
15.2.1.4 Establish testing and validation ownership	*272*
References	274

15.1 Understanding the economics of medical device cost

Many medical device manufacturers face the challenge of providing projected cost and time to develop the medical device. The cost of the device will require extensive planning and will depend on a range of factors such as follows:
- Development challenges and opportunities
- Quantifying market size
- Projected volumes

Table 15.1 Cost–product development cycle.

Cost type	Product development cycle
Material	Concept, design, development
Fabrication	Design, development
Validation	Validation
Quality	Entire product lifecycle

- Regulatory pathways
- Required testing

Table 15.1 shows relationship between cost and product development cycle. The overall cost of a device is dependent on the following:
1. Material
2. Fabrication
3. Quality
4. Validation

15.1.1 Material

Material selection for a new medical device can drive the cost of the finished good and impact entire product lifecycle, starting from product design through regulatory approvals and commercialization [1].

The selected materials should provide product design freedom and promote regulatory compliance, simplify manufacturing and appealing to providers and patients by enabling ergonomic, comfortable, and aesthetic solutions. To achieve commercial success in a shorter time, it is important to emphasize to meet these criteria's when selecting materials. Several criteria are considered when selecting materials including cost of the material [1].

Material costs are a critical aspect of any medical device. It is critical for manufacturers to look beyond the simple price of a plastic to assess its true lifecycle costs. Different materials can raise or reduce costs by affecting downstream factors, such as follows [1]:
- Amount of material required per device—This is influenced by device design, manufacturability, processing, material density, wall thickness, and part consolidation
- Material form, i.e., pellets, powder, and so on
- Additives required
- Secondary processing required
- Manufacturability

- Ease and speed of manufacturing due to required secondary operations or a material's suitability for high-volume production
- Transport costs for shipping weight or distance

It is important to consider how each candidate material can add or reduce the unit cost of a device throughout the full lifecycle. Sometimes a more expensive, higher-performing material can actually save money in the long run.

15.1.2 Fabrication

Determining the cost of fabrication depends on number of factors. While the factors affecting costs are varied, there are some major contributors that can be considered for creating reasonable estimates [2].

15.1.2.1 Material

In the device manufacturing process, material is one of the biggest factors in determining the costs of fabrication. Fabrication can be cheaper or expensive depending on the type of material that is used for fabrication. For example, specialty plastics would cost more than if engineering plastic was used for fabrication. In comparison with plastics, different metals will cost differently for fabrication. For example, steel is going to be more expensive than aluminum [2].

Additional factors that drive the cost are as follows:
- Part design such as the thickness of the material
- Labor cost
- Material properties
- Requirement for single or multiple materials

15.1.2.2 Labor

Labor is another factor that needs to be considered when determining the cost. The term labor is generally referred to both human and mechanical labor. Mechanical labor (machines) plays a supportive role to humans in a manufacturing process and is an important to factor in determining overall cost [2].

15.1.2.3 Part design and accuracy

Part design is also a factor in determining the cost of manufacturing a part. Less complex parts will require fewer machining steps such as cutting, milling, bending, and so on and therefore will cost less. Parts that are complex would require to undergo sequence of machining steps and therefore will cost more. Greater complexity simply translates to greater costs.

In addition to the part design, accuracy is another factor that will determine increase in the cost. Higher accuracy means that the part would spend more time being machined and therefore leads to increase in the cost [2].

15.1.2.4 Finishing and quantity

It is common for a part to undergo finishing step such as assembling, polishing, and painting. These additional steps would add to the costs and is dependent on the user needs and product requirements.

Similarly, it is important to understand that there will be increase in the total cost when the number of parts being produced is increased. However, with each additional reproduction, the average cost per item is going to decrease. Due to a number of different factors that then can include getting multiple parts from, a single piece of material and not having to set a machine up again for producing multiple copies [2].

15.1.3 Quality

McKinsey performed a market study to understand the cost of quality for medical devices. The study concluded that the direct cost of quality was approximately 6.8%—9.4% of industry sales. One-third of these costs came from the direct costs of ensuring good quality; the remaining two-thirds from the direct costs resulting from poor quality [3,4].

Quality of the medical device and the manufacturing process can drive the cost of the finished good. Costs that are generated from quality can be divided into three types [5]:
- Direct costs of ensuring good quality
- Direct costs of poor quality
- Indirect quality cost

15.1.3.1 Direct costs of ensuring good quality

Direct costs of ensuring good quality are the organizational costs of prevention and appraisal and represent the largest share of the total costs. McKinsey estimates that this cost is approximately around 2.0% to 2.5% and consists of the following activities [5]:
- Quality system support
- Validation
- Quality control
- Testing and inspection

- Auditing
- Other quality activites

Although all of these activities are primarily performed by quality organization, there is also a cross-functional contribution from operations organization. R&D and other organization do not contribute to these costs. The cost of prevention and appraisal varies by technology:
- 1.5%–2.0% of sales for disposables and implants
- 3.5% of sales for small electromechanical devices and capital equipment

15.1.3.2 Direct costs of poor quality

Direct costs of poor quality are associated with the following activities:
- Remediation
- Routine internal quality failures
- Routine external quality failures
- Nonroutine external quality failures

Remediation

Remediation costs account for 0.4%–0.7% of annual sales and involves a broad range of activities performed mainly by the quality organization. These activities include the following:
- Updating the design control documentation to ensure adherence to quality management system (QMS) standard, ISO 13485
- Reverification and revalidation of processes to ensure adherence to QMS standard, ISO 13485
- Performing investigations and identifying the root cause
- Corrective and preventive actions (CAPA)
- Complaints
- Medical device reports

Routine internal quality failures

Routine internal quality failures account for 2.1% of annual sales. Sources of these failure costs are as follows:
- Rejects and rework
- Deviations

Rejects and rework Depending on the medical device market, the cost associated with rejects and rework can vary between 20% for small electromechanical devices and 50% for implants and disposables.

Deviations Cost associated with deviation activities may involve the following:
- Deviation resulting from nonadherence to the procedures
- Deviation resulting from production quality failures:
  - Destroyed materials
  - Inventory changes
  - Fees for support

Routine external quality failures
Routine external quality failures account for 0.4% to 1.6% of annual sales. Sources of these failure costs are as follows:
- Warranty costs
- Returned and destroyed product costs
- Handling costs
- Administrative costs such as complaint intake and support for returns

Nonroutine external quality failures
Nonroutine external quality failures account for 1.9%–2.5% of annual sales. Sources of these failure costs are as follows:
- Product recalls
- FDA observation letter, Form 483
- FDA warning letter
- Consent decrees
- Import bans
- Consumer litigation

15.1.3.3 Indirect quality costs
Indirect quality costs are associated with the following activities:
- Revenue loss and market cap due to nonroutine quality failures
- Disproportionate cost as a result of potential compliance action

15.1.4 Validation

Validation activities are performed to provide objective evidence that the device being validated meets the design criteria and can consistently produces parts to its predetermined specifications using an established process. Determining the cost driven by validation depends on a number of factors. There are two types of validation activities that drive the cost:
- Design validation
- Process validation

15.1.4.1 Design validation
Design validation costs are driven by a number of factors [6]:
- Device design
- Prototyping
- Testing
- Engineering change order revisions

Device design
This activity starts at the design stage of the project and accounts for up to 80% of the product development cost. Making decision early in the design phase can significantly impact the cost of the product. The overall cost of development will also determine the end user cost. Keeping development costs down can ultimately increase margins while keeping prices competitive. Performing finite element analysis tool and prototypes early in the design phase will save time and money and can increase margins due to overall lower developmental costs [6].

Prototyping
This stage is used to validate that the product is error free and holds up to specifications through the manufacturing process. Sources that drive the cost of the product are as follows [6]:
- Time required to build the prototypes
- Number of prototypes iteration
- Capability to build prototypes

Reducing the number of prototypes in the design process by implementing design validation can significantly reduce the cost of the product.

Testing
Significant cost is associated with testing the product. Here, testing is performed on product and has to be retested when failure occurs. Outsourcing the testing can be costly, and even, bringing it in-house requires extensive capital cost. Additional costs are associated with time spent writing the protocol, performing the testing, and finalizing the report for number of studies per iteration of the product. Not to mention, developing fixtures to adequately represent loads as seen in the real world will add additional cost to the device final cost [6].

Engineering change order revisions

Engineering change orders (ECOs) are a result of analyzing the test data and determining that a change is required. Significant costs are associated with the release of the ECOs. Table 15.2 shows ECO processing timeline for change request. A reasonably complex change request comes in (medium change) for a component redesign. This change will take an average of 17 days to be completed from the time the request is made. Eliminating one ECO per project would save 3 weeks of time and consequently, save cost of the product [6].

15.1.4.2 Process Validation

Process validation costs are driven by a number of factors:
- Resource planning
- Personnel training
- Designing of qualification runs
- Testing
- Good documentation practices (GDP)

Resource planning

Significant cost is associated with inefficient resource planning. Resources are people, equipment, place, money, or anything else that is needed to do all planned activities. Inefficient planning could lead to increase in the overall cost. To assign resource to each task, it is important to identify resource availability. Assign resources that are available to perform the task.

Table 15.2 Engineering change order (ECO) processing timeline [6].

	# of days to process		# of ECOs processed monthly	
	Average	Range	Average	Range
Minor change (e.g., misc. Drawing correction, minor bill of materials fix)	3.7	0.1–24	34	2–100
Medium change (e.g., new SKU, component redesign)	16.9	1–70	24	2–100
Large change (e.g., new product line, product line design, large-scale cost reduction)	142.1	5–520	2.9	1–9

Personnel training
It is not enough to stress that personnel training is critical when it comes to performing qualification tasks. Performing a task without being trained to the equipment or the process can result in noncompliance to FDA regulation and consequently lead to heavy fines and significant added cost toward the end product.

Designing of qualification runs
Qualification runs should be planned and backed by statistical design of experiments. Performing a qualification run is a huge undertaking and require coordination from cross-functional team. The following are list of high-level steps:
1. Writing a protocol and operating procedure
2. Scheduling the trial
3. Allocating resources to run the trials
4. Procuring raw materials
5. Executing the run
6. Testing the product
7. Evaluating the results (C_{pk}, P_{pk})
8. Writing the report
9. Routing the report for approval using ECO

Each of these steps has huge compliance ramifications involved and can lead to significant increase in cost. Failing a qualification run has a huge impact on cost as well as the qualification process would likely have to be repeated that adds additional increase in the cost.

Testing
It is importance to emphasize selecting the appropriate sample size for testing the product based on the risk and criticality of the product. Selecting less sample for testing may lead to deviation, a risk of repeating the run, and impact on the overall cost. Selecting excessive samples for testing would increase the cost without added benefits.

Good documentation practices
GDPs are guidelines to record raw data entries in a legible, traceable, and reproducible manner. Incomplete documentation or falsifying the data has huge monetary and legal ramifications for medical device companies.

This not only lead to reperforming qualification runs but also can cause companies legal and settlement fees, loose revenue/reimbursements, federal prosecution, and huge fines from regulatory agencies.

15.2 Migration of device manufacturing, validation and use

The term "migration" here refers to transferring production from one manufacturing facility to another. The following factors may influence the final decision:
- External factors
- Internal factors

External factors occur independent of actions of existing manufacturing leadership. The list includes the following:
- Market trigger
- Cost savings
- Proximity to market

Internal factors are derived from existing relationship with either a contract manufacturer or internal operations team. The list includes the following:
- Quality concerns
- Delivery issues

15.2.1 Consideration for manufacturing transfer

15.2.1.1 Planning

Planning starts with a well-defined project plan document that sets the stage in the manufacturing transfer. Jabil has defined a manufacturing transfer checklist with list of key considerations that needs to be accomplished during the planning phase. Fig. 15.1 shows the manufacturing transfer checklist [7].

15.2.1.2 Consistent communication

Consistent communication is the second phase in the manufacturing transfer process. It is very important for a successful manufacturing transfer and it is during this phase that a comprehensive communication strategy is established. Fig. 15.2 shows the manufacturing transfer checklist for the communication phase as provided by Jabil [7]. Some of the proposed considerations are as follows:
- Defining responsibilities
- Determining "pitcher"/"catcher" responsibilities if ownership changes

MANUFACTURING TRANSFER CHECKLIST
PLANNING

Consideration / Action	Customer	Manufacturer
Assign experienced Project Managers (PMs) for both parties *Clearly define roles and responsibilities. PMs are the conductors and must be knowledgeable and empowered to run the program.*	X	X
Gather and share all product documentation	X	
Review product documentation to assure complete sets *Run a careful check. Watch for drawings within drawings, layers of detail, accurate part numbers, etc.*		X
Provide information on suppliers *Include the Approved Vendor List (AVL) and any updates to it. Outline special pricing and discount arrangements. Report on current inventory.*	X	
Inventory Assessment *Is there a need for buffer inventory? Are there supplies at customer that should be shipped to CM?*	X	X
Assess supply chain and part/product for End of Life (EOL) *Is buffer inventory available? Does current demand for part or product indicate EOL? Are any parts obsolete? Are there shortages? Are value-add engineering efforts required?*	X	X
Determine testing process *Who owns the testing process? If owned by third party, can we use them? What is cost? Include validation plan detailing the process and timing. Provide "good" products to confirm testing.*	X	
Re-validate testing process and equipment – ensure tests are repeatable and reproducible (R&R) prior to full production		X
Document Transfer Plan *Create a schedule to meet with weekly status calls (or more often). Allow for flexibility in plan (plan well, but expect the unexpected)*	X	X

Figure 15.1 Manufacturing transfer checklist—planning. *Source: Reproduced with permission from JABIL Healthcare.*

- Establishing timing commitments for responses
- Setting up weekly status calls with executive business reviews
- Identifying and communicating concerns at the earliest

15.2.1.3 Materials
During the manufacturing transfer process, it is important to have a clear understanding of the material flow, i.e., where materials are coming from

MANUFACTURING TRANSFER CHECKLIST
COMMUNICATION

Consideration / Action	Customer	Manufacturer
Define responsibilities for each partner *Determine focal point for each team member*	X	X
Determine "Pitcher / Catcher" responsibilities if ownership changes	X	X
Establish timing commitments for responses and hold to them	X	X
Set up weekly (or more often) status calls with quarterly executive business reviews	X	
Identify and communicate concerns ASAP – failure history, reliability, obsolescence, etc.	X	X

Figure 15.2 Manufacturing transfer checklist—communication. *Source: Reproduced with permission from JABIL Healthcare.*

and where they are going. Every material has to be accounted, and a detailed list of inventories of all required materials, including their costs, their lifecycle stages, and their supply chains, is prepared.

Fig. 15.3 shows manufacturing transfer checklist that outlines considerations for materials [7].

15.2.1.4 Establish testing and validation ownership

The last phase during the manufacturing transfer process is establishing testing and validation ownership. In this phase, test processes and procedures are created and verified such that they meet validation and compliance standards.

MANUFACTURING TRANSFER CHECKLIST
MATERIALS

Consideration / Action	Customer	Manufacturer
Create and share a detailed listing of suppliers, parts and materials. *Include AVL and any updates to the list. Are any parts obsolete or nearing EOL? Are the materials performing to expectations?*	X	
Determine pipeline for the materials	X	X
Transfer excess/existing inventory from supplier to CM	X	X
Address obsolete materials. *Transition Lifetime buys.*	X	
Identify and communicate underperforming materials. *Should value-added engineering (VAE) be applied to create a better performing material? If adding new, improved materials – plan new test and validation.*	X	X

Figure 15.3 Manufacturing transfer checklist—materials. *Source: Reproduced with permission from JABIL Healthcare.*

Fig. 15.4 shows manufacturing transfer checklist for testing and validation phase. Jabil has recommended the following considerations [7]:
- Ownership of the validation master plan
- Transferability of test equipment
- Identifying owners of test equipment
- Developing a plan while keeping revalidation in mind
- Having additional buffer in the schedule

MANUFACTURING TRANSFER CHECKLIST
TESTING & VALIDATION

Consideration / Action	Customer	Manufacturer
Determine who owns the validation plan	X	X
Can current test equipment be transferred? Does it need to be replaced? Does it need to be replicated? Build timing considerations for testing into plan.	X	X
Define who owns: In-circuit tester (ICT), Functional test systems and Fixtures	X	X
Provide "known good" units to validate testing	X	
Build into the plan and allow for sufficient time to re-validate the testing system to ensure results are repeatable and reproducible (R&R)	X	X
Consider buffer inventory needs based on testing and validation schedules	X	X

Figure 15.4 Manufacturing transfer checklist—testing and validation. *Source: Reproduced with permission from JABIL Healthcare.*

References

[1] A. Esposito, Ten Criteria for Choosing the Right Materials for Your Medical Device Design, Medical Design and Outsourcing, 2016. https://www.medicaldesignandoutsourcing.com/ten-criteria-choosing-right-materials-medical-device-design/.
[2] D. Trousil, Calculating Medical Device Manufacturing Costs, StarFish Medical. https://starfishmedical.com/blog/calculating-medical-device-manufacturing-costs/.
[3] E. Hoxey, "What is the cost of good quality for medical devices", Compliance Navigator for Medical Devices, The British Standard Institute, 2018. https://compliancenavigator.bsigroup.com/en/medicaldeviceblog/what-is-the-cost-of-good-quality-for-medical-devices/.
[4] Case for Quality, US Food and Drug Administration. https://www.fda.gov/medical-devices/quality-and-compliance-medical-devices/case-quality.

[5] T. Fuhr, E. Makarova, S. Silverman, V. Telpis, Capturing the value of good quality in medical devices, McKinsey and Company, 2017. https://www.mckinsey.com/industries/pharmaceuticals-and-medical-products/our-insights/capturing-the-value-of-good-quality-in-medical-devices#.
[6] 5-Ways Design Validation Will Save Money, Cadimensions. https://www.cadimensions.com/blog/5-ways-design-validation-will-save-money/.
[7] B. Barnhart, J. Vennari, Manufacturing Transfer Checklist—the Devils is in the Details, JABIL Healthcare. https://www.jabil.com/blog/manufacturing-transfer-checklist.html.

Index

Note: 'Page numbers followed by "f" indicate figures and "t" indicate tables.'

A

Abbott laboratories, 198
Acetal copolymer, 74
   applications, 109
   medical grades and suppliers, 109, 109t
   properties, 107–108, 108f, 108t
Acrylonitrile butadiene styrene (ABS), 74, 100–102
   applications, 102
   medical grades and suppliers, 102, 102t
   properties, 100–102, 101f, 101t
ALTUGLAS Rnew, 218
Anthrex Biocomposite, 214
Arista AH absorbable hemostat, 168
ARKEMA Rilsan polyamide 11 resin, 219
AROA bio endoform restorative bioscaffolds, 165–166
AROA bio myriad matrix, 166
AROA bio myriad morcells, 166
Automated external defibrillators (AEDs)
   Cardiac Science Corporation, 170
   Defibtech LLC, 172
   Philips medical systems, 172
   Zoll medical corporation, 170–172
Automate 2500 family, 200
Avitene hemostats, 168

B

Balloon distal thermal welding, 157
Balloon folding, 158
Balloon pleating, 158
Balloon proximal welding, 157
Biobased polymers
   ALTUGLAS Rnew, 218
   ARKEMA Rilsan polyamide 11 resin, 219
   diagnostics and labware devices, 218
   healthcare electronic devices, 218–219
   implantable medical devices, 210–218
      Anthrex Biocomposite, 214
      Biocomposites Inc. Bilok biocomposite, 216

      CONMED Linvatec GENESYS biocomposite, 215
      Eviva polysulfone, 216
      Evonik's biodegradable polymers, 210–212
      fused filament fabrication 3D-printed polyether ether ketone implant, 217
      Johnson and Johnson Biocryl Rapide biocomposite, 215
      MICRORAPTOR REGENESORB, 214
      RESOMER custom biodegradable polymers, 210–211
      Smith and Nephew REGENESORB, 213–214
      Solvay Solviva Biomaterials, 216–217
      Stryker Biosteon biocomposite, 216
      TephaFlex absorbable monofilament sutures, 218
      Veriva polyphenylsulfone, 216–217
      VESTAKEEP care grades, 212
      VESTAKEEP dental grades, 212, 213f
      VESTAKEEP i-grades, 211–212, 212f
      VESTAKEEP polyether ether ketone, 211–212
      VICRYL suture, 217
      Zeniva polyether ether ketone, 217
      Zimmer Biomet DuoSorb biocomposite, 215, 215f
   medical mask and medical tubing, 219
   Solvay Kalix 2000 series, 218–219, 218t
Biocompatibility, 66–67
Biocomposites Inc. Bilok biocomposite, 216
Bio-Rad laboratories Inc., 198–199, 199f
Biosurgery
   AROA Bio endoform restorative bioscaffolds, 165–166
   engineered extracellular matrix, 166
      AROA bio myriad matrix, 166
      AROA bio myriad morcells, 166
   hemostats, 167–168

Biosurgery (*Continued*)
　　Arista AH absorbable hemostat, 168
　　Avitene hemostats, 168
　reinforced bioscaffolds
　　Tela Bio OVITEX, 165
　　Tela Bio OVITEX PRS, 165
　sealants
　　Progel pleural thoracic sealant, 168
　　Tridyne vascular sealant, 169
　　Vistaseal fibrin sealant, 169
　tissue products, 166—167
　　dCELL technology, 167
　　DermaPure, 167
　　SurgiPure, 167
Bonding
　adhesive bonding, 140—141
　mechanical fastening, 139
　solvent bonding, 139—140
　ultrasonic welding, 140
　UV bonding, 140

C

Cardiac ablation catheter
　Medtronic's DiamondTemp ablation system, RealTemp, 174
　Medtronic's family of cardiac cryoablation catheters, 175
Cardiac pacemakers, 175, 176t—177t
Cardiac Science Corporation, 170
Cardiovascular heart rhythm
　automated external defibrillators (AEDs)
　　Cardiac Science Corporation, 170
　　Defibtech LLC, 172
　　Philips medical systems, 172
　　Zoll medical corporation, 170—172
　cardiac ablation catheter
　　Medtronic's DiamondTemp ablation system, RealTemp, 174
　　Medtronic's family of cardiac cryoablation catheters, 175
　cardiac pacemakers, 175, 176t—177t
　implantable cardioverter defibrillator (ICD), 175—178, 179t—180t
Catheter manufacturing process, 153—158
　balloon pleating and folding, 158
　balloon proximal and distal thermal welding, 157

　catheter shaft drawdown, 156—157
　extrusion process, 153—156
　manifold bonding, 158
　proximal and distal skiving, 157
　proximal fuse, 157—158
　RO marker swaging, 157
　secondary operations, 156—158
Catheter shaft drawdown, 156—157
Central Drugs and Standard Control Organization (CDSCO), 237
Ceramics, 49
Cleaning validation, 255—256
Computer numerical control
　lathes, 133—134
　machining, 130—134
　mills, 131—133
CONMED Linvatec GENESYS biocomposite, 215
Construction materials, 21
Contract manufacturer, 4
Corrective and preventive action (CAPA), 4
Cost of care, 17—18
Cost—product development cycle, 262, 262t
Cost reduction polymers, 20—22
Current good manufacturing practices (cGMP), 5

D

Danaher corporation, 199—200
　Automate 2500 family, 200
　DxA5000 total laboratory automation system, 200
　Power express laboratory automation system, 200
　Power link workcell, 200
Defibtech LLC, 172
Design validation, 267—268
Device classification panels, 33, 34t
Device design, 18—20, 19f, 253—254, 267
Device history file (DHF), 5
Device history record (DHR), 5
Device master record (DMR), 5
Devices availability, 20—22
Device system, 19—20
Die cutting, 139

Index

Dimensional stability, 69
Direct costs
   good quality, 264—265
   poor quality, 265—266
Distal skiving, 157
Document controls, 229
Drug flow path, 68
Dry heat sterilization, 61—63, 62t
   principles of, 62, 62t
   types of, 62—63
DxA5000 total laboratory automation system, 200

E

Elastomers, 46—47
Electric discharge machining (EDM), 134—136, 135f
Emergency use authorization (EUA), 35
Engineered extracellular matrix, 166
   AROA bio myriad matrix, 166
   AROA bio myriad morcells, 166
Engineering change orders (ECOs), 268
Ethylene oxide (EtO) sterilization, 60—61, 61t
   exposure, 61
   principles, 61
European Medicines Agency, 234—235
European Union Commission, 233—234
   medical device regulations, 234
Eviva polysulfone, 216
Evonik's biodegradable polymers, 210—212
Exemptions, 33—35, 34t
Expanded access, 5
Extrusion
   background, 148—150
   film extrusion, 150—151
   process, 153—156
   tubing extrusion, 150, 150f

F

Fabrication, 263—264
   finishing and quantity, 264
   labor, 263
   material, 263
   part design and accuracy, 263—264
Film extrusion, 150—151

Final microbicidal process validation, 256
Finished device, 5
Flared staking, 142
Flush staking, 143
Food and Drug Administration (FDA). *See* US Food and Drug Administration (FDA)
Forced-air type, 63
Fused filament fabrication 3D-printed polyether ether ketone implant, 217

G

Gamma radiation, 60
   sterilization process, 60
Glass, 49
   curing, 161
   formulation, 161
Good documentation practices (GDP), 5, 269—270
Graying populations, 16—17

H

Health Canada, 238
Healthcare electronic devices, 218—219
Hemostats, 167—168
   Arista AH absorbable hemostat, 168
   Avitene hemostats, 168
High-level disinfection, 254
High-performance demand plastics and elastomers
   acetal copolymer
      applications, 109
      medical grades and suppliers, 109, 109t
      properties, 107—108, 108f, 108t
   polydimethylsiloxane
      applications, 122
      medical grades and suppliers, 122, 122t
      properties, 121—122, 121f
   polyetheretherketone
      applications, 110
      medical grades and suppliers, 111, 111t
      properties, 109—110, 110f, 111t
   polyetherimide
      applications, 120—121
      medical grades and suppliers, 121, 121t
      properties, 119—120, 119f, 120t

High-performance demand plastics and
  elastomers (*Continued*)
  polyphenylene sulfide
    applications, 116, 117f
    medical grades and suppliers, 117, 117t
    properties, 115–116, 115f, 116t
  polyphenyl sulfone
    applications, 113
    medical grades and suppliers, 113, 113t
    properties, 112–113, 112f
  polysulfone, 113–115
    applications, 115
    medical grades and suppliers, 115, 115t
    properties, 113–114, 114f, 114t
  polyvinylidene fluoride
    applications, 119
    medical grades and suppliers, 119, 119t
    properties, 117–118, 118f, 118t
  thermoplastic elastomer
    applications, 125
    medical grades and suppliers, 125, 125t
    properties, 124, 125t
  thermoplastic polyurethane
    applications, 123–124
    medical grades and suppliers, 124, 124t
    properties, 122–123, 122f, 123t
High-performance polymers, 74–77
  acetal copolymer, 74
  polydimethylsiloxane, 76–77
  polyether ether ketone, 75
  polyetherimide, 76
  polyphenyl sulfide (PPS), 75–76
  polyphenyl sulfone, 75
  polysulfone, 75
  polyvinylidene fluoride, 76
  thermoplastic elastomer (TPE), 77
  thermoplastic polyurethane, 77
Hollow staking, 143
Hologic Inc.
  Panther fusion system, 201
  Panther link, 203
  Panther plus, 202
  Panther system, 201
Hospital cleaners, resistance to, 68
Hot knife cutting, 139

I
Implantable cardioverter defibrillator (ICD), 175–178, 179t–180t
Implantable medical devices, 210–218
  Anthrex Biocomposite, 214
  Biocomposites Inc. Bilok biocomposite, 216
  CONMED Linvatec GENESYS biocomposite, 215
  Eviva polysulfone, 216
  Evonik's biodegradable polymers, 210–212
  fused filament fabrication 3D-printed polyether ether ketone implant, 217
  Johnson and Johnson Biocryl Rapide biocomposite, 215
  MICRORAPTOR REGENESORB, 214
  RESOMER custom biodegradable polymers, 210–211
  Smith and Nephew REGENESORB, 213–214
  Solvay Solviva Biomaterials, 216–217
  Stryker Biosteon biocomposite, 216
  TephaFlex absorbable monofilament sutures, 218
  Veriva polyphenylsulfone, 216–217
  VESTAKEEP
    care grades, 212
    dental grades, 212, 213f
    i-grades, 211–212, 212f
    polyether ether ketone, 211–212
  VICRYL suture, 217
  Zeniva polyether ether ketone, 217
  Zimmer Biomet DuoSorb biocomposite, 215, 215f
Injection molding process, 151–153, 152f
Inspection, 247
International Medical Device Regulators Forum (IMDRF), 238–239
Interventional cardiology
  NC Trek coronary dilation catheter, 188, 189f
  SYNERGY bioabsorbable polymer stent, 187–188, 187f
  Trek and Mini Trek coronary dilation catheter, 188–189, 189f

Index

WATCHMAN left atrial appendage closure device, 185—187, 187f
XIENCE Alpine everolimus eluting coronary stent system, 190, 190f
XIENCE Sierra everolimus eluting coronary stent system, 188, 188f
Investigational device exemption (IDE), 5, 228
In vitro diagnostic (IVD), 6
　Abbott laboratories, 198
　Bio-Rad laboratories Inc., 198—199, 199f
　Danaher corporation, 199—200
　　Automate 2500 family, 200
　　DxA5000 total laboratory automation system, 200
　　Power express laboratory automation system, 200
　　Power link workcell, 200
　Hologic Inc.
　　Panther fusion system, 201
　　Panther link, 203
　　Panther plus, 202
　　Panther system, 201
ISO 9001, 6
ISO 13485, 6
ISO 14971, 6

J

Jetstream atherectomy system, 185, 186f
Johnson and Johnson Biocryl Rapide biocomposite, 215

K

Knurled staking, 143

L

Labeling, 256—257
　packaging controls, 231
　requirement, 231—232, 232t
Laser
　cutting, 139
　marking, 158
Low performance demand plastics and elastomers
　polyethylene
　　applications, 83, 84f
　　medical grades and suppliers, 82f, 84
　　properties, 82—83, 82f
　polypropylene
　　applications, 81, 81f
　　medical grades/suppliers, 82
　　properties, 80—81, 80f
　polystyrene
　　applications, 88
　　medical grades and suppliers, 88
　　properties, 87, 87f, 88t
　polyvinyl chloride (PVC)
　　applications, 86
　　medical grades and suppliers, 86, 86t
　　properties, 85, 85f, 85t—86t
Low-performance polymers
　polyethylene, 71
　polypropylene, 71
　polystyrene, 72
　polyvinyl chloride, 71
Lubrication, 69, 159

M

Manifold bonding, 158
Market trends
　medical devices, 244
　　industry, 243—244
　　market drivers, 243—244
　　market restraints, 244
　　medical technologies expansion, 244
　　safety and quality, 241—242
　　self-diagnosis/treatment, 244
Material costs, 20—21
Materials selection, construction
　biocompatibility, 49—56, 51f, 54t
　　data evaluation, 53—55
　　dose response, 55
　　exposure assessment, 55
　　hazard identification, 53—55
　　risk characterization, 55—56
　　toxicological risk assessment (TRA), 53—56
　ceramics, 49
　elastomers, 46—47
　extractables and leachables, 56—57
　glass, 49
　metals, 47, 48t
　plastics
　　improved comfort and safety, 46

Materials selection, construction (*Continued*)
  infection resistance, 46
  innovation, 46
  low cost, 46
  sterilization, 57–63
    dry heat sterilization, 61–63, 62t
    ethylene oxide sterilization, 60–61, 61t
    gamma radiation, 60
    steam (autoclave) sterilization, 58–60
  thermosets, 47
Medical device reporting (MDR), 6, 40, 232–233
Medical devices
  classification of, 28–33, 29t
    class I devices, 28–30, 30f
    class II devices, 31, 32f
    class III devices, 31–33, 32f
  cost–product development cycle, 262, 262t
  development timeline, 262t
  deviations, 266
  device classification panels, 33, 34t
  direct costs of good quality, 264–265
  direct costs of poor quality, 265–266
    nonroutine external quality failures, 266
    rejects and rework, 265
    remediation costs, 265
    routine external quality failures, 266
    routine internal quality failures, 265–266
  emergency use authorization (EUA), 35
  exemptions, 33–35, 34t
  fabrication, 263–264
    finishing and quantity, 264
    labor, 263
    material, 263
    part design and accuracy, 263–264
  FDA's accelerated approval program, 40–41
  FDA's breakthrough devices program, 41–42
    benefits, 41
    program features, 42
    program principles, 41–42
  global nature of, 15–16
  history of, 2
  indirect quality costs, 266
  industry, 243–244
  listing, 225
  market drivers, 243–244
  market restraints, 244
  market trends, 244
  materials, 7–11, 262–263
  metals, 7–9, 8f
  migration, 270–273
    consistent communication, 270–271, 272f
    materials, 271–272, 273f
    planning, 270
    testing and validation ownership, 272–273, 274f
  plastics
    applications, 13
    comfort and ease-of-use, 12–13
    cost, 13
    design flexibility, 13
    safety, 12
    sterilization, 12
  polymers, 9–11, 10f
  postapproval market requirements
    general requirements, 38–39
    medical device reporting, 40
    postapproval report, 39
    postmarket surveillance studies, 40
    premarket supplement, 39–40
  premarket approval, 36–38
  premarket notification, 510(k), 36, 37t
  regulations, 224, 227t
  regulatory requirements for, 223–240
  requirements of materials, 13–15
    preliminary assessment, 14
    rank, 15
    screening, 15
  socioeconomic factors, 16–22
    construction materials, 21
    cost of care, 17–18
    cost reduction polymers, 20–22
    device design, 18–20, 19f
    devices availability, 20–22
    device system, 19–20
    graying populations, 16–17
    material costs, 20–21

Index

processability, 20–21
product weight, 21
proliferation of procedures, 18
technologies, 22
user system, 19
US Food and Drug Administration (FDA), 27–28
validation
  design validation, 267–268
  device design, 267
  engineering change orders (ECOs), 268
  good documentation practices, 269–270
  personnel training, 269
  process validation, 268–270
  prototyping, 267
  qualification runs designing, 269
  resource planning, 268
  testing, 267, 269
Medical mask and medical tubing, 219
Medical technologies expansion, 244
Medium-performance demand plastics and elastomers
  acrylonitrile butadiene styrene, 100–102
    applications, 102
    medical grades and suppliers, 102, 102t
    properties, 100–102, 101f, 101t
  polybutylene terephthalate
    applications, 98, 98f
    medical grades and suppliers, 98, 98t
    properties, 97–98, 97f, 97t
  polycarbonate
    applications, 93, 94f
    medical grades and suppliers, 94, 94t
    properties, 92–93, 92f, 93t
  polymethyl methacrylate
    applications, 96, 96f
    medical grades and suppliers, 96, 96t
    properties, 95, 95f, 95t
  polyphenylene oxide
    applications, 100
    medical grades and suppliers, 100, 100t
    properties, 98–100
Medium-performance polymers
  acrylonitrile butadiene styrene (ABS), 74
  polybutylene terephthalate, 73

polycarbonate, 72
polymethyl methacrylate, 73
polyphenylene oxide (PPO), 73
Medtronic's DiamondTemp ablation system, RealTemp, 174
Medtronic's family of cardiac cryoablation catheters, 175
Metals, 7–9, 8f, 47, 48t, 160–161
  formulation, 161
  surface interaction, 161
MICRORAPTOR REGENESORB, 214
Migration, 270–273
  consistent communication, 270–271, 272f
  materials, 271–272, 273f
  planning, 270
  testing and validation ownership, 272–273, 274f

N

National Medical Products Administration (NMPA), China, 235–237
NC Trek coronary dilation catheter, 188, 189f
Nonconforming product, 230
Nonroutine external quality failures, 266

P

Panther fusion system, 201
Panther link, 203
Panther plus, 202
Panther system, 201
Personnel training, 269
Philips medical systems, 172
Plasma cutters, 134
Plastic medical device manufacturing processes
  catheter manufacturing process, 153–158
    balloon pleating and folding, 158
    balloon proximal and distal thermal welding, 157
    catheter shaft drawdown, 156–157
    extrusion process, 153–156
    manifold bonding, 158
    proximal and distal skiving, 157
    proximal fuse, 157–158

Plastic medical device manufacturing
  processes (*Continued*)
    RO marker swaging, 157
    secondary operations, 156—158
  extrusion
    background, 148—150
    film extrusion, 150—151
    tubing extrusion, 150, 150f
  glass
    curing, 161
    formulation, 161
  injection molding process, 151—153, 152f
  laser marking, 158
  lubrication, 159
  metals, 160—161
    formulation, 161
    surface interaction, 161
  plastics, 162
    formulation, 162
  siliconization, 159—162
    dispersion, 160
    material viscosity, 160
    moisture sensitivity, 160
    processing time, 160
    silicone, 159—160
    surface interaction, 160
Plastics, 162
  applications, 13
  comfort and ease-of-use, 12—13
  cost, 13
  design flexibility, 13
  formulation, 162
  global sourcing of
    assessment process, 247—248
    distribution, 248
    establish quantification, 246
    inspection, 247
    prequalification, 245—248
    procurement method, 246
    product information, 247
    product specifications, 246
    purchasing, 247—248
    quality assurance (QA), 245—248
    receipt and storage, 248
    receive product information, 246
    screen product information, 247
    solicit information, 246
    submission procedure, 246
  improved comfort and safety, 46
  infection resistance, 46
  innovation, 46
  low cost, 46
  safety, 12
  sterilization, 12
Plastics and elastomers
  FDA class I devices, 70—77
    high-performance polymers, 74—77
    low-performance polymers, 70—72
    medium-performance polymers, 72—74
  performance-based selection
    biocompatibility, 66—67
    conductive, 69
    dimensional stability, 69
    drug flow path, 68
    hospital cleaners, resistance to, 68
    lubrication, 69
    mechanical properties, 69
    radiopacity, 69
    ranking, 70
    screening, 66—69
    sterilization compatibility, 68
    translation, 66
Plastics fabrication techniques
  bonding
    adhesive bonding, 140—141
    mechanical fastening, 139
    solvent bonding, 139—140
    ultrasonic welding, 140
    UV bonding, 140
  machining of, 129—139, 131t
    computer numerical control lathes, 133—134
    computer numerical control machining, 130—134
    computer numerical control mills, 131—133
    die cutting, 139
    electric discharge machining (EDM), 134—136, 135f
    hot knife cutting, 139
    laser cutting, 139
    plasma cutters, 134

Index

sawing, 138
turning, 136—137
water jet cutting, 137—138, 138f
staking
   flared staking, 142
   flush staking, 143
   hollow staking, 143
   knurled staking, 143
   spherical staking, 142
   thermal staking, 142
   types of, 142—143
two-part molding, 143—144
Polybutylene terephthalate, 73
   applications, 98, 98f
   medical grades and suppliers, 98, 98t
   properties, 97—98, 97f, 97t
Polycarbonate, 72
   applications, 93, 94f
   medical grades and suppliers, 94, 94t
   properties, 92—93, 92f, 93t
Polydimethylsiloxane, 76—77
   applications, 122
   medical grades and suppliers, 122, 122t
   properties, 121—122, 121f
Polyether ether ketone, 75
Polyetheretherketone
   applications, 110
   medical grades and suppliers, 111, 111t
   properties, 109—110, 110f, 111t
Polyetherimide, 76
   applications, 120—121
   medical grades and suppliers, 121, 121t
   properties, 119—120, 119f, 120t
Polyethylene, 71
   applications, 83, 84f
   medical grades and suppliers, 82f, 84
   properties, 82—83, 82f
Polymers, 9—11, 10f
Polymethyl methacrylate, 73
   applications, 96, 96f
   medical grades and suppliers, 96, 96t
   properties, 95, 95f, 95t
Polyphenylene oxide (PPO), 73
   applications, 100
   medical grades and suppliers, 100, 100t
   properties, 98—100
Polyphenylene sulfide
   applications, 116, 117f
   medical grades and suppliers, 117, 117t
   properties, 115—116, 115f, 116t
Polyphenyl sulfide (PPS), 75—76
Polyphenyl sulfone, 75
   applications, 113
   medical grades and suppliers, 113, 113t
   properties, 112—113, 112f
Polypropylene, 71
   applications, 81, 81f
   medical grades/suppliers, 82
   properties, 80—81, 80f
Polystyrene, 72
   applications, 88
   medical grades and suppliers, 88
   properties, 87, 87f, 88t
Polysulfone, 75, 113—115
   applications, 115
   medical grades and suppliers, 115, 115t
   properties, 113—114, 114f, 114t
Polyvinyl chloride (PVC), 71
   applications, 86
   medical grades and suppliers, 86, 86t
   properties, 85, 85f, 85t—86t
Polyvinylidene fluoride, 76
   applications, 119
   medical grades and suppliers, 119, 119t
   properties, 117—118, 118f, 118t
Postapproval market requirements
   general requirements, 38—39
   medical device reporting, 40
   postapproval report, 39
   postmarket surveillance studies, 40
   premarket supplement, 39—40
Postapproval report, 39
Postmarket surveillance (PMS), 6, 40
Power express laboratory automation system, 200
Power link workcell, 200
Premarket approval (PMA), 6, 36—38, 226—227
Premarket notification, 510(k), 36, 37t, 226
Premarket supplement, 39—40
Prequalification, 245—248
Processability, 20—21
Process validation, 268—270

Procurement method, 246
Product
  information, 247
  specifications, 246
  weight, 21
Proliferation of procedures, 18
Prosthetic implants
  Johnson and Johnson, 194
  Otto Bock, 194–197
    prosthetic foot solution, 196
    prosthetic hip solutions, 197
    prosthetic knee solution, 195
    prosthetic socket solution, 196
  Smith and Nephew, 197
  Stryker corporation, 193
Prototyping, 267
Proximal fuse, 157–158
Proximal skiving, 157

Q

Quality assurance (QA), 245–248
Quality audit, 6
Quality system
  regulation, 6, 228–231
  requirements, 229

R

Radiopacity, 69
Ranking, 70
Reinforced bioscaffolds
  Tela Bio OVITEX, 165
  Tela Bio OVITEX PRS, 165
Remediation costs, 265
Reprocessing
  device design, 253–254
  factors affecting quality, 253–254
  FDA guidance on, 252–253
  high-level disinfection, 254
  methodology, 254
  reprocessing methodology, 254
  single-use devices (SUDs), 258
  sterilization instructions, 254
  validating cleaning, 254
RESOMER custom biodegradable
    polymers, 210–211
Resource planning, 268
Reusable medical devices

cleaning validation, 255–256
final microbicidal process validation, 256
labeling, 256–257
overview, 251–252
reprocessing
  device design, 253–254
  factors affecting quality, 253–254
  FDA guidance on, 252–253
  high-level disinfection, 254
  methodology, 254
  single-use devices (SUDs), 258
  sterilization instructions, 254
  validating cleaning, 254
RO marker swaging, 157
Routine external quality failures, 266
Routine internal quality failures, 265–266

S

Sawing, 138
Screening, plastics and elastomers, 66–69
Sealants
  Progel pleural thoracic sealant, 168
  Tridyne vascular sealant, 169
  Vistaseal fibrin sealant, 169
Self-diagnosis/treatment, 244
Siliconization, 159–162
  dispersion, 160
  material viscosity, 160
  moisture sensitivity, 160
  processing time, 160
  silicone, 159–160
  surface interaction, 160
Single-use devices (SUDs), 258
Smith and Nephew REGENESORB,
    213–214
Socioeconomic factors
  construction materials, 21
  cost of care, 17–18
  cost reduction polymers, 20–22
  device design, 18–20, 19f
  devices availability, 20–22
  device system, 19–20
  graying populations, 16–17
  material costs, 20–21
  processability, 20–21
  product weight, 21
  proliferation of procedures, 18

Index

technologies, 22
user system, 19
Solicit information, 246
Solvay Kalix 2000 series, 218–219, 218t
Solvay Solviva Biomaterials, 216–217
Spherical staking, 142
Staking
   flared staking, 142
   flush staking, 143
   hollow staking, 143
   knurled staking, 143
   spherical staking, 142
   thermal staking, 142
   types of, 142–143
Static-air type, 62
Steam contact, 59
Steam (autoclave) sterilization, 58–60
   benefits of, 58
   common mistakes, 59–60
   drying, 58
   moisture, 59
   pressure, 59
   process, 58–59
   steam contact, 59
   temperature, 59
   time, 59
Sterilization, 57–63
   compatibility, 68
   dry heat sterilization, 61–63, 62t
   ethylene oxide sterilization, 60–61, 61t
   gamma radiation, 60
   instructions, 254
   steam (autoclave) sterilization, 58–60
Stryker Biosteon biocomposite, 216
Submission procedure, 246
SYNERGY bioabsorbable polymer stent, 187–188, 187f

T

Technical file, 7
Temperature, 59
TephaFlex absorbable monofilament sutures, 218
Thermal staking, 142
Thermoplastic elastomer (TPE), 77
   applications, 125
   medical grades and suppliers, 125, 125t

   properties, 124, 125t
Thermoplastic polyurethane, 77
   applications, 123–124
   medical grades and suppliers, 124, 124t
   properties, 122–123, 122f, 123t
Thermosets, 47
Thoracic drainage systems, 190, 192f
Time, 59
Tissue products, 166–167
   dCELL technology, 167
   DermaPure, 167
   SurgiPure, 167
Toxicological risk assessment (TRA), 53–56
Traceability, 230
Translation, plastics and elastomers, 66
Trek and Mini Trek coronary dilation catheter, 188–189, 189f
Tubing extrusion, 150, 150f
Turning, 136–137

U

Unique device identifier (UDI), 7
User system, 19
US Food and Drug Administration (FDA), 27–28
   accelerated approval program, 40–41
   acceptance activities, 230
   breakthrough devices program, 41–42
      benefits, 41
      program features, 42
      program principles, 41–42
   corrective and preventive actions, 230
   design control, 229
   distribution, 231
   document controls, 229
   establishment registration, 225
   guidance on, 252–253
   handling, 231
   identification, 230
   installation, 231
   investigational device exemption (IDE), 228
   labeling
      packaging controls, 231
      requirement, 231–232, 232t
   medical device

US Food and Drug Administration (FDA) (*Continued*)
  listing, 225
  regulations, 224, 227t
  medical device reporting (MDR), 232—233
  nonconforming product, 230
  premarket approval, 226—227
  premarket notification 510(k), 226
  production and process controls, 230
  purchasing controls, 229
  quality system
    regulation, 228—231
    requirements, 229
  records, 231
  service, 231
  statistical techniques, 231
  storage, 231
  traceability, 230

V

Validation, 7
  cleaning, 254
  design validation, 267—268
  device design, 267
  engineering change orders (ECOs), 268
  good documentation practices, 269—270
  personnel training, 269
  process validation, 268—270
  prototyping, 267
  qualification runs designing, 269
  resource planning, 268
  testing, 267, 269
Vascular surgery
  Jetstream atherectomy system, 185, 186f
  tunneling solutions, 182—184

SLIDER graft deployment system, 184, 185f
vascular graft tunneling instrumentation, 182—184, 184f
vascular grafts
  Advanta VXT vascular graft, 181—182, 182f
  Flixene AV Access Grafts, 178—181, 181f
vascular patches, 182, 183f
Veriva polyphenylsulfone, 216—217
VESTAKEEP
  care grades, 212
  dental grades, 212, 213f
  i-grades, 211—212, 212f
  polyether ether ketone, 211—212
VICRYL suture, 217

W

WATCHMAN left atrial appendage closure device, 185—187, 187f
Water jet cutting, 137—138, 138f

X

XIENCE Alpine everolimus eluting coronary stent system, 190, 190f
XIENCE Sierra everolimus eluting coronary stent system, 188, 188f

Z

Zeniva polyether ether ketone, 217
Zimmer Biomet DuoSorb biocomposite, 215, 215f
Zoll medical corporation, 170—172

Printed in the United States
by Baker & Taylor Publisher Services